Gene Dreams

Gene Dreams

WALL STREET,
ACADEMIA,
AND THE RISE OF
BIOTECHNOLOGY

Robert Teitelman

BasicBooks
A Division of HarperCollinsPublishers

Library of Congress Cataloging-in-Publication Data
Teitelman, Robert.
    Gene dreams: wall street, academia, and the rise of biotechnology
    / Robert Teitelman.
        p. cm.
    Includes index.
    ISBN 0–465–02659–1 (cloth)
    ISBN 0–465–02658–3 (paper)
    1. Genetic engineering industry.   2. Biotechnology industries.
I. Title.
HD9999. G452T44   1989
338. 4'766065—dc20   89–42528
                                                          CIP

To my parents,

Raymond and Edythe Teitelman

"Truth," says Bacon, "comes more easily out of error than of confusion": but the view that I have to recommend is that confusion, *ignoratio elenchi,* is itself the most fatal of errors, and that it occurs whenever argument or inference passes from one world of experience to another.

—MICHAEL OAKESHOTT,
Experience and Its Modes

CONTENTS

Contents

PREFACE

THIS BOOK was not born from the brow of any Zeus; rather, it accreted, like an archaeological site, over a number of years. It began in the early 1980s when *Forbes* magazine dispatched me to write on cancer research. This was a tangled subject upon which neither *Forbes* nor I had much experience; the stories that resulted, I suspect, confused nearly everyone. Indeed, I was a scientific neophyte, as one inevitably is in the many specialized and arcane disciplines of modern science and technology; and yet even with the extreme complexity of modern science, of cancer research, it is a mistake, I think, to assume that reasonably intelligent observers— whether readers or writers—cannot penetrate the broad outlines and major issues. Even then, when oncogenes were just peaking as a public phenomenon, it seemed apparent that there was a widening gulf between the rhetoric of cancer research and the reality. Once one turned to the business issues, which inevitably meant biotechnology, a similar problem appeared: in this case, it manifested itself as a disparity between stock prices and the fundamental commercial potential of most of these companies.

That was the merest glimmering of the idea that motivated the writing of this book: that biotechnology, as a "revolution," has failed. Since then, a succession of kindly editors has afforded the time and space to allow this hypothesis to ripen. First, Peter Hall and then Geoffrey Smith at *Financial World* encouraged me to do a series of stories on major pharmaceutical companies, which provided an often-ignored perspective. And at *Oncology Times,* Debra Lumpe offered me a monthly column that gave me the chance to explore more recondite aspects of this subject, at a level of detail impossible in the financial press. I was, in hindsight, remarkably lucky: I have had, for the most part, editors who were not a part of the prevailing tendency toward blind optimism when it came to technological issues, notably Steve Kindel at *Forbes* and Geoff Smith at *Financial World.*

That, I think, is important. In all the talk that rattles on about science

Preface

writing, little is said about the framework that determines the way science is treated in the press. There are three prevailing, and self-reinforcing determinants: simplicity, optimism, and the notion that scientific subjects exist in a non-contextual world. Science and technology are last bastions of the High Victorian notion of progress; and thus their role, particularly in biology, is to provide miracles and hope to an eager public. This is not just the sin of the press. It is a gentle conspiracy shared by the scientific bureaucracy and self-interested academics and businessmen—that is, everyone who needs to raise money. This combination of crystallized ideology and the economic motive combine to create the overheated talk of transformation and revolution—or by the small but noisy reaction against science and technology, of disaster and annihilation—that so often accompanies these subjects.

This book attempts to thread a path between these extremes. It labors to make clear the importance of key distinctions: between the science—molecular biology—and the technology that it has thrown up; between long-wave Schumpeterian economic effects and the shorter-term commercial insurrection; between science performed in academia, and technology that drives the commercial sphere. Much of the hype, positive and negative, that has been generated by this highly publicized industry, I would argue, comes from confusing these distinctions, from believing that behavior suitable in one world is suitable for another.

I have already mentioned a few editors who deserve my thanks. I should not fail to mention Martha Millard, who began the whole process, or Steve Fraser, my editor at Basic Books, who not only quite literally saved this book but pressed me into creating a sturdier framework to place it in. There are many other individuals, as well, who aided and abetted these efforts, mostly through reading, commenting, and arguing over the manuscript; they are, needless to say, in no way responsible for my opinions: Stelios Papadopoulos, Wayne Fritzsche, Lenore Freeman, Macy and Toivo Koehler, Walter Gilbert, Chris Miles, Bob Mecoy. I am particularly grateful to Herbert Burkholz, who provided necessary doses of bracing advice. And, above all, to Camilla, who was always there.

Gene Dreams

The Mythos
of Biotechnology

THE SHELF ABOVE ME sags with volumes published over the past decade or so about this phenomenon called "biotechnology." So before proceeding, allow me to sketch out what this book does *not* aim to do. *Gene Dreams* is not a primer on biotechnology, describing, with all the requisite arrows and diagrams, twelve ways to clone a gene or recipes for generating monoclonal antibodies. It is not a novelistic look at life inside a biological laboratory or a warning about biological hazards. And it is certainly not a detailed history in the manner of Horace Freeland Judson's magisterial work on molecular biology, *The Eighth Day of Creation*;[1] both biotechnology and I need a bit more seasoning before attempting that. What I have tried to do here is more modest: to begin an exploration of why biotechnology blossomed so furiously in the early 1980s, why it has taken on the characteristic form that it has, and where it is going from here.

Many of the themes of this book were motivated by questions that arose in reporting on one discrete aspect of the biotechnology industry or another. I chose as the central case Genetic Systems, because it seemed to illuminate more aspects of the industry than any other. Why, I wondered, did academic biology suddenly begin to disgorge hundreds of its practitioners to companies in the late 1970s? Why was Wall Street so ready to embrace these fledgling companies, such a long way from generating products, and then so quick to abandon them just a few

3

GENE DREAMS

years later? Was biotechnology comparable to the electronics revolution? How has biotechnology's umbilical cord to Wall Street determined research programs and product development? What changes has it wrought on the way America supports biomedical research?

Despite its youth, the reality of biotechnology is often lost within a fog of its own enthralling mythology. Unlike the more cloistered world from which it sprang, biotech is a fully public—even a political—phenomenon. It is an industry dependent on money raised by Wall Street from nonscientific investors: it has products and personalities that have to be promoted and sold; and it is molded by implacable financial pressures. Like colonists, it has tried to take root in a landscape and culture very different from its birthplace.

Biotechnology—the very word was invented on Wall Street—is a set of techniques, or tools, not a pure science like much of academic biology. As a set of tools, it can be used to advance scientific experiments or to produce viable, useful products such as human pharmaceuticals and diagnostics. Technology, by definition, interlocks with cogs and wheels of the workaday world. My purpose is to show, as graphically as possible, how that interaction between academic biology and commercial technology evolved, and the consequences that resulted.

The industry, of course, has a much grander image of itself than just another industrial enterprise. Particularly in its early days, biotech entrepreneurs talked as if they were assembling academic laboratories that happened to be funded by Wall Street instead of some nonprofit or governmental body. They argued that they could combine both the pure science of the academic world and the product development of the drug industry; that they could excel as both scientists and inspired entrepreneurs; that armed with these powerful tools and their own inspiration, they could, like some romantic hero, transcend the realities of corporate life.

That kind of thinking was mirrored in the way these firms were portrayed to the outside world. In the popular press, the industrial realities and limits were minimized. The technology burned so brightly that it seemed to erase finer distinctions. Biotechnology was thus viewed as either Promethean, engaged in saving humankind, or Faustian, dabbling with devilish forces beyond our ken. Interferon as a cancer cure was the flip side of nightmarish visions of Andromedan strains conjured up by critics like Jeremy Rifkin.[2] Even the financial press and the Wall Street research community followed the inner logic of these caricatures. Here, the focus tended to be on scientific wonders—on scientific *possibilities*—while skipping quickly across the more immediate, more

4

contingent, technical, social, and business realities. A company that has mortgaged its future or lacks development skills or has no money left or has a product no one wants faces a limited future, no matter how powerful its technology. In the pursuit of truth, science needs no market; technology, on the other hand, has no reality beyond its application and exists only in relation to the marketplace. Biotechnology, however, was regularly viewed as a science story, cast in the future tense, safe from realistic analysis. And, in truth, because Wall Street would pay for the future, such dreams could, occasionally, become self-fulfilling. Bad news, as they say on Wall Street, helps no one but the short-sellers. Good news sells.

At the root of some of this was a messianic hope that still stubbornly clings to the technology. The grand, and currently unrealistic, hope for a simple, easy, comprehensive cancer cure was just a part of this; it was also the belief that a new, fundamental technology had a sort of alchemical ability to transform all that it touched—from the physical being of humanity to the national economy. This is nothing new. Successive dreams of secular technological utopias, based on steam, rails, atomic energy, electricity and electronics, and now biology, have long been part of the American experience. America began as a dream of a lost, secular Eden. And Americans have often fantasized of a return to Eden through technology—in literary critic Leo Marx's words, "the machine in the garden."[3] Indeed, less than twenty years ago, James Carey and John Quirk could write, in an essay called "The Mythos of the Electronic Revolution," of the mentality that seizes on such technological breakthroughs:

> The futurist mentality has much in common with the outlook of the Industrial Revolution, which was heralded by the Enlightenment philosophers and nineteenth century moralists as the vehicle for a general progress, moral as well as material. Contemporary images of the future also echo the promise of an eighth day and thus predict a radical discontinuity from past history and the present human condition. The dawn of this new era is alternately termed the "post-industrial society," "post-civilization," and "the global village." The new breed of man inhabiting the future is characterized as the "post-modern man," the "protean personality," the "post-literate electronic man."[4]

Similar claims have been made for biotechnology. The title of Judson's history, for instance, refers to the eighth day of redemption following the apocalypse. And Gunther Stent, a researcher at the University of California at Berkeley and one of the pioneers of molecular biology, elaborated on the theme in his 1969 book, *The Coming of the Golden Age: A View of the End of Progress,* in which he described the discovery and manipulation of DNA as signaling an end to social and economic evolution.[5] As one observer commented, DNA would provide the kind of material well-being that Marx and other utopians believed achievable through the Industrial Revolution. History had ended; Eden could be recreated on earth. Faustian man was being gradually phased out of the environment he had created.[6]

More prosaically, there was that ubiquitous catchphrase of biotechnology, "magic bullets," which referred, however loosely, to the therapeutic dream of creating pharmaceutical regimens that would strike only at the center of a disease, not at the healthy tissue around it. All drugs have some side effects, however minor; and cancer therapeutics—radiation and chemotherapy—are among the least focused. Thus, to speak of magic bullets was to present the possibility of a transcendent leap in efficacy, a way to talk about a *cure* without mentioning that word. Indeed, whenever anyone spoke of the possibilities of biotechnology, the words "magic bullets" invariably crept into the conversation.

"Magic bullets" was a compact and memorable phrase. Even better, it was open-ended: it could take on a multitude of meanings. At first, just monoclonal antibodies fit beneath its garish, if narrow, marquee. After all, the German chemist Paul Ehrlich had identified antibodies as potential "magic bullets" when he coined the phrase nearly a century ago. But once biotechnology was off and running, the phrase "magic bullets" quickly spread across the landscape. One of the first manifestations of this process was the transmutation of the phrase to "silver bullets," a harmless enough change that replaced Mr. Wizard with the Lone Ranger. But the alteration went considerably further. If monoclonal antibodies were magic bullets, why not say the same of exotic substances such as interferon? Doesn't interferon strike at some cells, say tumor cells, leaving normal cells unharmed? *In theory,* doesn't that make interferon a magic bullet? And if interferon was a magic bullet, what of targeted drugs produced by other means? By 1980 the phrase had cast off its ropes to the solid deck of science and drifted into the warm southern seas of publicity. A magic bullet was more a declaration of intent, like a utopian vision of the future, than anything to do with the practical application of science to medicine.

This messianism fused with the enthusiastic gospel of the entrepreneur. If the technology was so powerful, was not a new form of business

The Mythos of Biotechnology

demanded—a postindustrial corporation? And, if so, would not the grinding day-to-day affairs of business—financing, management, payroll, accounts receivable—become irrelevant? Those were the kinds of notions that led some executives and observers to underestimate the business issues, a situation analogous to a long jumper dismissing the effects of friction, gravity, and a strong breeze. The creative deal loomed over the careful husbanding of resources; the role of personality took precedence over structure and systems; the technological breakthrough was expected to blow aside all barriers. Biotechnology, went the notion, is a revolutionary technology wielded by entrepreneurial companies engaged in a transformation of the economic landscape. The stagnant world of American big business, particularly those dinosaurs known as the drug companies, are doomed; a new age is dawning; a new dispensation has arrived.

Reality proved more intractable. Commercializing products that actually make money turned out not to be simple at all. Many failed. Meanwhile, the drug companies did not collapse; in fact, through product licensing, acquisitions, joint ventures, and huge in-house investments in biotechnological techniques, they are recouping aggressively. And diseases like cancer have not given up their secrets easily. Thus, although the industry has carved itself a niche in the biomedical research world, there is no guarantee that it will retain it. The technology, now over a decade old, has spread so far and is now so common, that the notion of a separate biotechnology industry has begun to blur along the edges.

Early on, a biotechnology company was defined as one that specialized in either gene splicing or monoclonal antibodies. But now those two techniques are common to the armamentarium of any biological laboratory. Perhaps, then, a biotech company is one specializing in biologicals rather than in organic chemicals. But again, most firms have accepted the necessity to work with more complex molecules like biologicals. Perhaps, then, a biotech company is an entrepreneurial, development-stage company pursuing classic molecular biology—an inductive, rather than empirical, approach. Fine, but that severely underestimates the capabilities of the drug companies and relegates the biotechnology industry to a world of small research boutiques. In fact, what separates the successful biotechnology firm from a traditional drug company? Is Genentech a biotech company, a pharmaceutical company, or something new—a biopharmaceutical company?

One cannot doubt the historic importance of biotechnology as a scientific and commercial set of tools. After all, it has already had a major impact on drug development, disease diagnostics, and agriculture; and as time goes on, those tools will alter both products and markets and force a variety of knotty ethical questions. Perhaps biotechnology, allied with information processing, will provide the fuel for a Schumpeterian long wave of economic development.[7] The central question involves, however, the translation of this technology into social or business structures. What structures, what changes, will biotechnology create over the long term? Will biotechnology toss up an entirely new, self-sustaining industry, like the large computer and semiconductor companies spawned by the electronics revolution? Or will these powerful tools be subsumed into structures and technologies that already exist? Will biotechnology, in fact, create permanent changes in the ways we fund basic biomedical research? Is biotechnology a revolutionary or an evolutionary force? This book, in short, is really about limits: the limits of both the entrepreneur and the technology.

The key lies with the word *revolution*. Like magic bullets, revolution is a concept that dons many guises and gets used for many purposes. In the early days of the industry, this notion often took on an insurrectionary sense: the conviction that this upsurge in entrepreneurial firms would "overthrow" the large, conservative pharmaceutical companies that sat atop commercial biomedicine. But that was not all. More quietly, it subverted the other major fiefdoms of the biomedical establishment, including the system of academic research and the enormous generation of research funding that came, for the most part, from the federal government through the distributive mechanism of the National Institutes of Health (NIH).

At the very least, biotechnology represented a radical decentralization of the biomedical establishment. By opening up new avenues of financing, biotechnology offered a way of getting around the established peer-review system and the dominance of institutions like the NIH. Although it was a practical outcome rather than a stated goal of biotechnology, this decentralization garnered applause and popular support. It spoke to a widespread public sense of disappointment in the progress of the heavily publicized War on Cancer begun by President Nixon in 1971; indeed, crazes like Laetrile, the quack cancer cure made from apricot pits, was fueled by the belief that there was a government conspiracy to block a cancer cure. It was arguably no coincidence that Laetrile faded only when interferon came onto the scene in 1980. The War on Cancer raised public expectations but offered no real sense of participation; biotechnology tapped that public directly

8

The Mythos of Biotechnology

for support. Access to that new and bountiful financing source, with all its deceptive independence from traditional constraints, also spawned other consequences. Many of the most visible phenomena of the biotechnology era—the hype, the wild claims, the disappointment—arose directly from this decentralization of decisionmaking, this sudden opening up of decisionmaking to the invisible hand of the market, and from the scientific ignorance that underlay it. And like the War on Cancer, biotechnology created its own long-term problems: it raised expectations to an impossible level while spawning its own competition, as hundreds of entrepreneurs rushed in to join the bonanza.

Genetic Systems, the company examined most closely here, offers a window on this process. A small Seattle company assembled by a pair of young Wall Street deal makers and run by a young scientific entrepreneur, Genetic Systems struggled to succeed in a difficult environment—costs were higher, competition tougher, and the science sketchier than anyone thought. To many observers, the 1985 acquisition of Genetic Systems by Bristol-Myers (and, simultaneously, of a somewhat similar company called Hybritech by drugmaker Eli Lilly) marked the end of the heroic age of biotechnology, a loss of innocence, despite the profits some investors took home. And today that does seem to have been the case. There have been successes—particularly the approval of tissue-plasminogen activator, a drug that dissolves blood clots, developed by Genentech, long considered the premier biotechnology company—but they have not been on the heroic scale that the early biotech dreamers and promoters foresaw. And each success, in turn, has been shadowed by failure. Even Genentech has had to struggle with the Food and Drug Administration, patent litigation, a failure to get Medicare reimbursement, mounting competition, and a market for tissue-plasminogen activator that has simply not been anywhere as large as the company and Wall Street expected. Similarly, the elegant theories that created so much excitement in the early days of the industry, with talk of magic bullets and immunotherapy, have proven to be full of hidden complexities and unfortunate contingencies. That is not to say they are not fertile scientific theories, only that their components have not yet proven themselves as technologies.

But this, of course, is nothing new. Judson, for instance, rounds out his history of molecular biology in 1970, when researchers had begun to believe that they understood the general operation of the cell, at least in the simplest creatures. "Many molecular biologists were con-

9

fident," he writes, "that the outline could be stretched to take care of higher organisms. They spoke of filling in classical molecular biology." But, he continues, "The confidence was premature. The problem was harder, both technically and conceptually, than anyone had forseen. Some of the difficulties could not possibly have been predicted. Life turned out to be full of surprises."[8]

CHAPTER 1

The Biomedical Triangle

O N MOST DAYS, vast tides of capital surge through Wall Street, sweeping over feeding beds of investors and corporations alike. The composition changes continuously, with new companies appearing as the old wither and die, or shuffle off into merger or acquisition. Most initial public offerings—that is, the first sales of stock of a company to the public—involve scores of investors, lawyers, and investment bankers but fail to have much of an impact on the greater community of Wall Street. But on October 14, 1980, a much larger event was in the works: a company from south San Francisco named Genentech was coming out. Genentech called itself a biotechnology company—a new word, a new idea—and seized the promiscuous imagination of Wall Street. The company was the creation of two men, a biochemist named Herbert Boyer and a young venture capitalist turned entrepreneur named Robert Swanson. Boyer, with another colleague, had been the first to pluck a gene from one organism and insert it into another successfully. This was the first example of a deliberate genetic recombination, although the process had already acquired, like a Hindu deity, many names: genetic engineering, recombinant DNA, gene splicing, *biotechnology*. The company had already won a fierce race between two academic laboratories to use these intricate tools to insert a gene for human insulin in bacteria; it had since licensed that work to Eli Lilly, the world's largest seller of porcine, or

11

young pig, insulin to diabetics, for commercialization.[1] And then there was *interferon*—another new, endlessly fascinating name like some mythical African kingdom. Interferon was a powerful natural protein that was reputed to combat everything from herpes to cancer to the common cold.

All this put Genentech into a different category from other new companies. Genentech's substance was highly complex, even esoteric. Most investors deal with complexity as if it were a tax problem: they try either to hide or shelter it, or they dump it, like a shoebox crammed with receipts, on an expert. With interferon, even the experts were undecided. "Just the bottom line!" investors demanded. Alas, there was no bottom line; there would not be one for years. Into that vacuum stepped the brokers, a breed that, in a world of slamming phones, learn how to be concise. Brokers are salespeople, and nothing sells like hope, the service economy's answer to the widget. Interferon embodied hope and the promise of fabulous future profits. *Time* magazine had run interferon on its cover earlier in the year ("The Big IF"), and *Time* was about as technical as most Wall Streeters ever got.

The brokers found a receptive audience. Genentech had a definite allure—a blend of technology, management, and class. "By owning Genentech," said one money manager, "you were the envy of everyone in the conversation pit." Besides, in an economy wracked by inflation, beset by imports, confused by years of zigzagging economic policies, a decade in which stocks had skidded so low that one business magazine predicted, wrongly, the death of the stock market, "technology" looked like the closest thing to salvation short of Japan falling into the sea. Wall Street was loaded with investors who dreamed of another wave of profitable new investment vehicles like the computer and semiconductor companies or, before that, the Polaroids and Xeroxes. Not that Genentech's prospectus did not warn about risk. Genentech had not yet produced a single product; human insulin would not be fully tested and approved until 1984, interferon several years later. It had no experience in the drug business. It would have to raise considerably more money in the future.

No matter. The enthusiasm ran so high that pricing proved tricky. How great would be the demand? Genentech, with its lead underwriters, Blythe Eastman Paine Webber in New York and Hambrecht & Quist in San Francisco, elected to sell a million shares at thirty-five dollars each. This was quite high for a small, untested company. The fever spread swiftly. The financial fundamentals—the multiples, the projections of sales and earnings, the esoteric math by which Wall

The Biomedical Triangle

Street judges risk—got trampled as anticipation built. Thirty minutes after the market opened, the underwriters had sold their stock and disbanded; the twenty-seven market makers, through which trading was now funneled, absorbed the brunt of the storm. Twenty minutes later the now freely trading stock hit *eighty-nine dollars* a share. For the rest of the day it rose and fell, tumbling as investors cashed in their profits, spinning upwards again on new speculative enthusiasm. In midafternoon, John Whitehead, a managing partner at the brokerage firm of Goldman Sachs, came over the Dow Jones news wire expressing concern at the speculative fever. The trading slowed, then accelerated again. Late in the afternoon the stock took a final run into the upper eighties, only to fall back to seventy on the closing bell.

That eighty-nine-dollar high set a record for an initial public offering. The financial pages the next day recounted classic American tales of instant wealth, including one about a California Institute of Technology graduate student who suddenly discovered he had become, overnight, a paper millionaire. Several years earlier he had done some lab work for Genentech and had received 30,000 shares; when he left the company, soon after, he returned half of them.

The Genentech offering sent a shock through other small, private biotech operations and their venture-capital investors and investment bankers. In the months that followed, dozens of similar companies, many with equally beguiling names and comparable products, lined up to offer stock. Several succeeded in raising more money than Genentech. A firm called Cetus, for instance, with three Nobel laureates on its scientific advisory board, raised $120 million, almost drowning the market in shares. It was an impressive outpouring of capital. Even more unusual was the number of academics involved in these new firms. This was a new phenomenon: never had a new industry arisen with university scientists playing such a major role. In just a few years there would be over 100 public biotechnology companies fueled by some $500 million in new publically invested capital. The phenomenon soon took on a name: biomania.

The Genentech offering, like a stone tossed in a still pond, would for years generate ripples in the world of biomedical research. It produced an exhilarating break with the past and seemed to symbolize a major shift in the power structure presiding over both basic research and its technological component, pharmaceutical and diagnostic development. It brought to the game new players in the form of Wall Street investment bankers, brokers, and analysts and, by extension, a larger, scientifically unsophisticated public. And it altered not only the rules of the game but the kind of projects that were financed and pursued. Genentech, in essence, was the first round in a torrid tango of mutual seduction between Wall Street and a new class of biological entrepreneur. By 1981 many researchers began to think of Wall Street not as a grim Golgotha, but as a Golconda—in the words of financial historian John Brooks, a legendary city where "everyone who passed through got rich."[2] And for Wall Street operatives and their customers, Genentech and biotechnology offered the promise of medical miracles, big profits, and a plethora of deals.

But to understand truly what a radical departure Genentech was, one has to understand the institutional complexity that dominated biomedical research. A decade before Genentech, biomedical research in the United States had been swept by another sea change: the War on Cancer legislation, which poured billions of dollars into basic biological research. Yet President Nixon's 1971 decision to seek a cure for cancer was essentially conservative. It buttressed, for a time, the dominant role played by the federal government in biomedical research and fixed bureaucratic relationships, making them appear to have assumed a natural and inevitable division of labor. In fact, the particular constellation of institutions that ruled over biomedical research, like a variety of other spheres dominated by the federal government, had only developed in the years after World War II. The War on Cancer not only expanded an already large constituency that existed to disburse and spend that money; it also created an ever-increasing need, as the field expanded to consume the funding and as institutions labored both to retain and expand their prerogatives.

Before Genentech, biomedical research was dominated by three interrelated realms. The first was that of the pure or basic researcher. This was a world of graduate students, postdoctoral fellows, and laboratory chiefs at private research institutions such as Rockefeller University or Cold Spring Harbor Biological Laboratories, at universities like MIT or Harvard, or at federal laboratory complexes such as the National Institutes of Health. This realm involved teaching and training, as well as research. Basic biological research was just that: an exploration into fundamental mechanisms of biological structures. Its vanguard was made up of investigators into a murky molecular universe, seekers after truth without the taint of the commercial and political world. They were searching for biological mechanisms: What causes a cell to proliferate in a cancerous fashion? How does the immune system operate? Why does blood clot? Why do arterial vessels clog? Increas-

ingly, their efforts were driven by molecular biology, the highly sophisticated dissection of the molecular watchworks of the cell that burst
into public consciousness with Watson and Crick's 1953 elucidation of
DNA's double helix. Molecular biology studied the inner workings of
the cell, particularly the controlling mechanism, DNA.

Academic researchers coexisted uneasily with a second locus of
power, the large drug and chemical companies. Here, applied science,
or technology, ruled. Although Merck or Eli Lilly or Squibb funded
some basic research, these companies really focused on discovering
active compounds and turning them into useful therapeutics. In theory,
the pharmaceutical establishment would create new, commercially viable products based on foundations laid down by academic researchers.
In actuality, the two realms often seemed to work in separate com-
partments, communicating, only occasionally, by tapping on the walls.
Compared to the computer industry, biology was still fumbling in the
dark. There were, as yet, few applications of fundamental breakthroughs. "Biologists work very close to the frontier between bewilder-
ment and understanding," said Nobel laureate Sir Peter Medawar in
1968. "Biology is complex, messy and richly various, like real life; it
travels faster nowadays than physics or chemistry (which is good because it has further to go), and it travels nearer the ground."[3]

The drug companies were not interested in science without applications. Their approach was predominately empirical, relying on what
worked, without worrying overly about cause.[4] For years, the industry
had developed new products by shuffling through as many compounds
as they could lay their hands on. Pith-helmeted explorers would return
from exotic lands laden with baggies of dirt, plants, or mold scraped off
bark, boot soles, or the scummy surface of brackish bogs. Back at the
lab, scientists would test the samples against a variety of microbes and
viruses growing in row after row of carefully labelled petri dishes. If a
compound proved effective, an organic chemist—*not* a biologist—would
fiddle with it, snipping off a carbon atom here, adding a sugar ring
there, and then screen all over again, testing all the time. Is it toxic?
How much effect does it have? What if you try this or that? Petri dishes
were everywhere, and drug laboratories resembled giant cupcake
factories.

It was, at its heart, a bureaucratic science: an attempt to organize
and rationalize serendipity. The more compounds screened, the better
the odds of finding the new penicillin or the new cephalosporin, or
discovering something that worked. Explanations could be devised later.
The process sometimes took decades. Lederle Laboratories spent fif-

teen years screening and testing before stumbling over ethambutol, an antitubercular. As late as 1985, Squibb introduced a new antibiotic called Azactam that had first been found in 1978 in the Pine Barrens of New Jersey.

Scientifically, mass screening lacked precision and elegance; industrially, it worked, although by the late 1970s it was no longer working as well as it had. The number of breakthrough drugs, compounds that, by their very success, would create whole new markets, was slowing; the industry was awash with "me too" compounds that acted in very similar ways. The wars against old diseases, microbial infections like tuberculosis or pneumonia, were now triumphs of the past; and, indeed, the patents on those drugs had long since lapsed. The challenge was now to conquer forbidding problems like cancer, heart disease, arthritis, and Alzheimer's disease. Nonetheless, in pharmaceuticals, genius appeared far less frequently than the need to report earnings. Perspiration, the theory went, could make up for anemic inspiration. And throughout the 1970s most drug companies clung to the old ways: mass screening and organic chemistry.

The third realm of the biomedical establishment was a bureaucracy that focused more on politics and the allocation of resources than on actual research. Biological research was expensive and time-consuming, particularly in academia, where there were no profits to feed back to the laboratory. Thus arose a bureaucracy of fund raisers, grant givers, administrators, lobbyists, and politicians. This extensive network mushroomed in the years after World War II when Americans, suddenly rich and powerful, became grimly serious about good health. Before the war, research funding had been scarce. The National Institutes of Health (NIH), for example, traces its history to 1887 and a one-room bacteriological laboratory in a government marine institute. In 1937 when the National Cancer Institute (NCI) was opened, its budget was only a few hundred thousand dollars.[5] Still, by 1950, Congress, prodded by massive grassroots fund raising from such new organizations as the American Cancer Society, the American Heart Association, and the March of Dimes (which saw its first cause effectively eliminated when Jonas Salk developed the first polio vaccine in 1957) took up medical research in a big way. Birth-to-death health care was quickly becoming a right rather than a privilege. Organizations took the first steps to involve a larger public actively through mass fund raising drives. The response, in turn, had its effect on Congress and the administration. Who could argue with the thousands of women marching door to door with their cans jangling with nickles and dimes? Who could argue with an issue that had so much pork barrel and so little risk?

By the late 1970s the federal government dominated the funding of basic biological research. Overall, it was pouring 11 percent of all federally funded research-and-development (R&D) monies into basic biomedical research, compared to 2 to 4 percent for most other developed countries. As a result, Americans in 1979 published 40 percent of the world's articles on biology and 43 percent on biomedicine, and they dominated the Nobel Prizes for physiology or medicine. The division between basic and applied research showed up in these spending patterns as well. Unlike pharmaceutical R&D, federal funds supported a large educational establishment. Thus, while the numbers of basic researchers such as molecular biologists, immunologists, and biochemists soared in the late 1970s, relative shortages developed in microbiology, bacteriology, and bioprocess engineering, usually considered the applied, or industrial, end of biology.

As an institution, the National Institutes of Health profited the most from this continually expanding pool of funds. By the 1970s the NIH had eleven separate institutes specializing in everything from allergies to heart disease to cancer. The largest, by far, was the National Cancer Institute, which by 1981 was spending $989 million dollars a year on cancer research alone. Like all the NIH facilities, NCI served as both a world-class laboratory and a powerful funneling mechanism for funding other researchers and other laboratories, some $783 million in 1981.[6] The flood of War on Cancer funding made NCI the metropolitan center of cancer research, feeding laboratories located on an increasingly dependent periphery, either at the nation's twenty-two designated cancer research centers—centers such as Memorial Sloan-Kettering in New York, M. D. Anderson in Houston, or Fred Hutchinson Cancer Research Center in Seattle—or at a myriad of academic centers. At the same time, NCI succeeded in attracting a stable of great talent, making its own "intramural" laboratory a powerful and prestigious force. It thus acquired enormous influence: it could effectively determine the direction of basic biological research through its control of grants, and it could define the cutting edge through the work of its own scientists. Biology in general responded like a plant to the shining sun of NCI's largesse. Researchers who had once plied byways like developmental biology or virology soon learned to write their grants in a cancer context. Inevitably, NCI defined a kind of acceptable research which, not surprisingly, resembled the molecular biology that its own leading lights practiced.

GENE DREAMS

*   *   *

This triangulation of power contained the seeds of rebellion or, at least, competition. By the early 1970s, molecular biology was growing increasingly confident in its own achievements. "By 1970, a coherent outline of the processes of life had been put together," concluded H. F. Judson.[7] If it proved to be not quite as coherent as first imagined, that did not undermine the almost imperial confidence of the discipline. Throughout the previous two decades, the field had progressively cleared more ground around the elegant, shining insight of Watson and Crick: that the spiral staircase of DNA was a perfect molecular structure for storing and replicating genetic information. The basic mechanism of genetic inheritance had been elucidated; the interplay of RNA and DNA in the production of cellular proteins, Francis Crick's so-called central dogma, sketched out; bits and pieces of metabolic pathways uncovered; the structure of a few major proteins laboriously puzzled out; and a thousand other small parts of a vast, complex puzzle nailed down. In cancer research, many believed they were closing in on a mechanism through their work with tumor-causing viruses, an area NCI pioneered and encouraged.

Still, the drug industry seemed to react sluggishly. Molecular biology was striking as an intellectual exercise, but like cosmology or astrophysics, it did not provide a treatment for that child dying of a rare inherited cancer, nor had it spun off a new generation of profitable and useful drugs. For all the Nobel Prizes, mechanisms manipulable by an organic chemist or a pharmacologist remained elusive. And, except in a few exceptional cases, the drug companies remained aloof.

Then, in the mid-1970s came the two now-legendary breakthroughs that changed everything. The first resulted from a collaboration between Herbert Boyer at the University of California at San Francisco and Stanley Cohen at Stanford. With a basic understanding of the role of DNA in cellular processes mapped out, researchers had begun talking of altering genes in order to alter the production of cellular proteins. Boyer and Cohen provided the technique that made this molecular splicing work. Boyer had been experimenting with a so-called restriction enzyme which could slice a double strand of DNA at a fixed spot, leaving a single strand dangling behind like a strap on a bus. Slicing up pieces of DNA yielded a tool for mixing and matching lengths of different pieces of DNA. Cohen had been working with plasmids, tiny rings of DNA just a few genes long that float outside the main chromosomes in bacteria. The experiment, now done routinely in undergraduate laboratories, was a marvel of ingenuity. The pair used Boyer's enzyme to cut Cohen's plasmid and insert a piece of foreign DNA. The plasmid then carried the foreign DNA into the bacteria. But would the new genes function within the bacteria, ordering up foreign proteins? And would they be replicated when the bacteria divided? In both cases the answer was yes, and genetic recombination, or recombinant DNA, was born.

The hazardous implications of recombinant DNA were immediately apparent (although how real they are remains debatable). The products were, after all, foreboding Frankensteinian cells. As a result of those fears, molecular biology struggled throughout the seventies with the safety issue, suffering through moratoriums, protests, and strict guidelines about the use of recombinant DNA—another reason the highly conservative drug companies shied away.

The second fundamental scientific breakthrough took place far more quietly. It involved antibodies, an important part of the immune system that identifies and attacks foreign threats such as viruses or bacteria. Again, a collaboration occurred. In 1975, Georges Köhler, formerly of the Basel Institute for Immunology, arrived at the Medical Research Council Laboratory of Molecular Biology (MRC) in Cambridge, England, to study mutations in genes that order specific antibodies. He worked under Cesar Milstein, a leading expert on the immune system. There was, alas, a major obstacle. Generating pure quantities of any particular antibody proved all but impossible. To get antibody against, say, a flu virus, a researcher would inject the virus into a bleating sheep. The sheep would produce antibodies against the virus—generically known as an antigen—but it would also make antibodies against lots of other things that together would constitute the sheep's resistance to disease. This conglomeration was known as a polyclonal mix. For Köhler to pin down which mutation created which antibody, he needed cells to produce pure lots of antibodies—in the jargon, monoclonal antibodies.

Milstein had recently done some experiments in which he fused rat and mouse cells, producing a sort of hybrid cell, or a hybridoma. That triggered what Köhler called his "crazy idea." Why not take cells that produce antibodies—called B-lymphocytes—and fuse them with myeloma cells, cancer cells that were "immortalized," that would grow and divide indefinitely. To Köhler's surprise, his little experiment worked. The B-lymphocytes contributed the antibody-producing genes, while the cancer cells drove their rapid proliferation. The resulting hybridoma would pump out quantities of monoclonal antibodies.

Few immediately recognized the importance of what seemed, at first, to be a laboratory trick. Milstein wrote a note to his MRC bosses suggesting that there might be industrial applications, including a patent, for the invention. They ignored it. Milstein and Köhler then submitted a short paper to *Nature*, the British scientific journal, but the editors buried it among the letters describing baboon behavior and goldfish physiology. Soon the pair were back studying antibodies, now and again shipping interested researchers tiny samples of their hybridoma through the mail, a common scientific practice. Only then did the implications of the discovery begin to be felt: researchers now had the means to produce large amounts of very specific antibodies for a variety of experiments and, perhaps, for therapy. Köhler and Milstein shared the Nobel Prize in 1984.

What linked these two discoveries? How could they trigger such a massive shift beneath the placid surface of the research establishment? Both were tools rather than profound insights into fundamental biological mechanisms. Recombinant DNA and monoclonal antibodies blurred the strict division—and, to many researchers, suddenly narrowed the gap—between basic and applied biology. All kinds of practical possibilities now began to suggest themselves, particularly against cancer. And practical applications inevitably suggested the possibility of venturing forth into the strange and forbidding world of business.

CHAPTER 2

Chiefs, Postdocs, and Entrepreneurs

In this society, you're made to feel stupid if you can't make money.
—CESAR MILSTEIN

IN THE LATE 1970s, three trends were beginning to coalesce in this expansionary world of molecular biology. Lab chiefs were running increasingly larger organizations; a critical mass of postdocs had developed; and in the business world, the technological entrepreneurs had come into their own. The three groups shared a certain ethos: All felt as if they were ideological outsiders, working against the stultification of conventional wisdom and conventional institutions. And each believed in the power of science and technology.

By the mid-1970s, molecular biology was booming. The flood of War on Cancer money built labs, funded teaching programs, and created the need for more money. Armies of bright young recruits poured into the ranks, and labs groaned with postdocs, poorly paid but highly skilled young scientists who had completed their graduate education but who still required seasoning—biology's answer to the intern. For many postdocs it was a disturbing time: all this excitement and all this thrill of discovery, recombinant DNA, hybridomas, interferon. But it was also increasingly competitive: academic slots were low paying and in great demand, and there never seemed to be enough money. Scientific fraud seemed to be increasing.[1] Like the inflation of the period, demand for

21

funding outran the historically plentiful supply. Many postdocs were just entering their best years—like athletes, biologists seem to peak as bench experimentalists in their early thirties—and they still had to worry about grants. It didn't seem fair. The cancer program, the source of most of the funding, was rife with rumors of budget cuts, or in the jargon of the bureaucrat, rebudgetizing, reprioritizing. The postdocs, a nervous, fretful lot, knew who would lose out in the event of cuts—and it wouldn't be the sixty-year-old Nobel Prize–winning boss.

Add to this unstable mixture of frustration and ambition the entrepreneur. The motherland of the entrepreneur was Silicon Valley and the technology was electronics. Silicon Valley extended roughly from the Stanford campus in Palo Alto, just south of San Francisco, down Route 101, through Cupertino and Menlo Park and Santa Clara to San Jose on the dry eastern lip of the county. Once a land of farms and orchards, the valley was no longer particularly scenic. The highway rolled past one fabricated, windowless box after another, sitting amidst acres upon acres of blacktop parking lots. The names on those buildings, however, were gold on Wall Street: Hewlett-Packard, Intel, Fairchild, Apple Computer. No matter that more high technology had grown up around Los Angeles, 400 miles to the south, or along the corridor stretching from Boston to Washington, D.C., Silicon Valley was all that high technology *plus* entrepreneurs.

By the late 1970s the economy was stalling. Inflation was up, trade was down, and insecurities were rampant. Who was to blame? Big business and big government, the same villains of the sixties. The widely suggested antidote was the entrepreneur, particularly the kind that came out of Silicon Valley. These entrepreneurs carried the aggressive idealism of the 1960s into the new calling of business. They possessed the "entrepreneurial spirit," which could transform the individual while reviving the economy. The entrepreneur radiated, literally or figuratively, youth; and the entrepreneurial startup, with its creativity and flexibility, was often contrasted with the arthritic stiffness of larger, older organizations. On Wall Street, the paradigmatic entrepreneur was Apple Computer's cofounder, Steven Jobs, who, in the mid-1970s had just emerged from his garage, proselytizing about the manifest destiny of the microcomputer. "I'm just a guy who should have been a semi-talented poet on the Left Bank," said Jobs once. "I got sidetracked here."[2] In 1983, Jobs was worth $210 million in Apple stock. He was rich and young; he had an artistic glaze; in fact, he seemed to have it all.

Entrepreneuring as a full-time calling—and many in Silicon Valley, from busboys to CEOs, thought of it that way—requires someone else's money to make it work. An entrepreneur begins with an idea. He then seeks investors, from rich relations to venture capitalists. If he successfully piques their interest, he trades his concept for cash. For his backers, success hinges on the public offering; this is their escape hatch. A public offering creates a public market, where shares can be sold at will. That transforms an illiquid investment into liquidity, or cash, which, depending on the glamour of the company, magically multiplies the original stake. Stock bought for pennies may now sell for dollars. Ten times, fifty times, a hundred times the original investment is not unusual. Penny stock in Genentech, for instance, produced 350 times the original stake on the opening. If you had unloaded at the top on that first, furious day, you could have made 890 times your original stake; if you had held on, you could have gotten even more as the stock rose, split, rose, and split again. This kind of mathematics drove venture investing, although it did not always work quite that well.

In this entrepreneurial world, the venture capitalist occupies an ambivalent position. Like a gigolo, he's involved, but not involved. He's part entrepreneur, part accountant. He's Santa Claus *and* Ebenezer Scrooge. One thing is clear, however: It is easier to get rich in venture capital than in entrepreneuring. It is also less risky, though hardly risk free, which is why so many entrepreneurs became venture capitalists. One Apple Computer, one Tandem, one Digital Equipment, one Genentech can cover a multitude of venture-capital duds, just as one hit movie can pay for a dozen flops. And the really smart venture capitalists launch themselves with someone else's money.

By 1975 the Silicon Valley entrepreneur was in full bloom. Semiconductors had spawned electronics; electronics, in turn, had spawned computers, large and small, plus disc-drive firms, software developers, and terminal makers. The personal-computer boom was just getting started. Meanwhile, a few began to sense a new trend, in Wall Street terms, a new high-tech concept. One of these was Robert Swanson. Stocky, balding, with premature jowls that *Barron's* once compared to Mickey Rooney's,[3] Swanson in 1976 was twenty-seven, a junior partner at Kleiner Perkins, a venture-capital firm with offices on Sand Hill Road, close to Stanford's Technology Park, where Silicon Valley had been born.

Kleiner Perkins had invested some money in a small startup called Cetus. Swanson was assigned to oversee it. Cetus had been founded in 1971, before Boyer and Cohen's recombinant breakthrough, to exploit molecular biology. Its two founders, Ronald Cape, a biochemist, and Peter Farley, a physician, both with business degrees, hooked up with Donald Glaser, who had won a Nobel Prize in physics in 1960 before switching to molecular biology, a relatively common evolution. Cape and Farley were classic Silicon Valley entrepreneurs. "We're not a recombinant DNA company per se," Cape said once in a fit of enthusiasm. "We're not a hybridoma company per se, we're not a mass screening company per se, and we're not an industrialization and fermentation company. We're all of these."[4] The media dubbed Cape, like Jobs, a "visionary," and *Fortune* magazine anointed him "the poet laureate of biotechnology."[5] Farley was blunter than his partner, although equally optimistic. "We're building," he said modestly, "the next IBM."[6] Indeed, Cetus dabbled in everything from genetically engineered bacteria for alcohol production to antibiotics and human proteins to new vaccines—all encouraged by investors such as Standard Oil of Indiana, Standard Oil of California, National Distillers, and Shell. These were the days when oil was synonymous with endless reserves of money.

The venture capitalists at Kleiner Perkins were not convinced. Entrepreneurs were fine, as long as they showed a little strategic discipline. Besides, Swanson wanted his own show. So in 1976 the partnership sold out their stake in Cetus to the Canadian minerals company Inco. Around that time, Swanson, checking into the commercial possibilities of recombinant DNA, put in a call to Herbert Boyer at UCSF; he did not, at the time, realize Boyer's central role in genetic engineering. Could he come up to talk? Boyer was busy but said sure, if you can say your piece in twenty minutes. On a Friday afternoon, Swanson drove over to the green confines of UCSF. He appeared in the lab attired not, like Boyer, in faded jeans and vest, but in a suit and tie; the grad students and postdocs kibitzing in the hallway thought that humorous. Who *is* this guy? Boyer, burly and bearded, closed his door; a few minutes later, the two came out. "I'll be back later," said Boyer. "I'm getting a beer."

The short interview grew longer. "Can recombinant DNA be commercialized?" asked Swanson. Boyer was enthusiastic. He even had a name for a company: genetic engineering technology, or Gen–EN–tech, with the accent on the second syllable. Swanson and Boyer chipped in $500 each; Kleiner Perkins then came through with $100,000 or so. Swanson had a plan. Undoubtedly an entrepreneur technically, Swanson lacked the sheer excess of the breed. He did not rant and rave like Jobs; he did not engage in visionary monologues like Cape; he did not stage rock concerts, disappear into the woods for weeks, or recklessly speed around in foreign cars. Still, he knew entrepreneurs had to loosen up sometimes to recruit and keep the young researchers that Boyer, who remained at UCSF, was helping to recruit.[7] Swanson thus injected an academic atmosphere into a commercial framework. For instance, he instituted Ho-Ho's, Friday afternoon beer bashes, in which he often dressed up in a funny costume like a grass skirt or a bumblebee outfit. Ho-Ho's were common in Silicon Valley, but they soon became part of the Genentech image. Look, these guys are *human*. They give out stock options to employees. They allow their scientists to publish.[8] The CEO dresses like a *bumblebee*. That style, in turn, created an army of admirers. *Esquire* once went so far as to call Swanson not only the father of bioindustry—which was inaccurate—but the inventor of something called "post-industrial management," which was simply meaningless. Somehow, said *Esquire*, Swanson had created a new form of business organization fitting for such a revolutionary technology.[9]

Swanson's real assets were more mundane, and perhaps more rare, than that. He recognized that the entrepreneur and the postdoc had much in common, and he worked to cement that bond. He knew how to make a business operate, how to finance it sensibly, and how to create and use publicity. For years, Genentech enjoyed a remarkable run of good publicity. Swanson himself was often photographed in rolled-up sleeves or a white lab jacket, peering at the dials of an impressive-looking fermenter. But beneath the white lab jacket, he invariably wore his red corporate tie.[10] Being in charge meant hiring employees to fiddle with the dials. And unlike most entrepreneurs, who eschew discipline in favor of spontaneity, Swanson, after some early forays into animal products and diagnostics, developed a clear plan. He allowed no dabbling in energy projects for big oil companies, and no research into areas just because they seemed neat; instead, he insisted that all efforts be applied to a single, overarching goal of developing human pharmaceuticals through recombinant DNA techniques—marketable, potentially mind-bogglingly profitable human biopharmaceuticals. Swanson wanted products.

Wall Street, for the most part, was not interested yet in the drama taking place at startup companies like Genentech. If it was not a public company—if one could not invest in it—Wall Street pretended it did not exist. Slowly, however, word seeped out. In 1979, Nelson Schneider, a pharmaceuticals analyst for the brokerage firm of EF Hutton in New York City, began to feel the rumbling. Normally, Schneider analyzed the major drug stocks, writing reports and recommending that stock be bought and sold. One day he noticed that Eli Lilly was buying some technology for producing a human form of insulin from a South San Francisco company called Genentech. Lilly was one of the great old names in American pharmaceuticals and was a company he followed closely. This news struck him as a bit odd; Lilly traditionally preferred to keep its research at home. Four months later, Schneider was in London for a conference on biomedical research. Several speakers talked about new techniques such as recombinant DNA and mentioned several more small companies. Back in New York, Schneider read about another big drug company, Schering-Plough, investing in a company called Biogen, based in Cambridge, Massachusetts, and Geneva, Switzerland. And clients were beginning to ask what was going on. What *was* recombinant DNA anyway? So in August 1979 he sat down and wrote a report entitled "DNA: The Genetic Revolution" and passed it on to clients. It was essentially a technology primer, heavy on the optimism.

The report evoked such a reaction that Schneider took the next step: he decided to hold a conference. He drew up a speaker list: Bob Swanson, Peter Farley of Cetus, Phillip Sharp of Biogen, Leslie Glick of Genex, and Zolt Harsanyi, a researcher assembling a report on genetic engineering for the Congressional Office of Technology Assessment. It was the 1979 who's who of genetic engineering. He booked a small room at the Plaza Hotel and issued the invitations, expecting anywhere from 35 to 100 people—a normal crowd for this sort of meeting. But in the days just before the meeting, he had to change the room four times; 500 money managers, investors, investment bankers, and analysts eventually crowded into the hotel to hear about this new phenomenon. "It was not planned," said Schneider. "It happened. Not one of these companies was public. People in the audience began asking: How do we invest in this thing? Where do we put our money? I know that Bob Swanson sat there and saw the money sitting there on the table at the Plaza."[11]

CHAPTER 3

Raising
Consciousness

Interferon is a substance you rub on stockbrokers.
—A scientist quoted in *Forbes*, September 1980

THE GENENTECH OFFERING
was a sign that a new way of doing business had fully emerged. But to
understand fully the inner dynamics of biomania, one has to turn to
interferon, the wonder drug that fueled not only Genentech but so
many other early biotech offerings. The protein interferon was discovered, studied, and nurtured under the traditional establishment—it was
embraced, in the beginning, only by a small, if vocal, segment of the
biomedical bureaucracy—but it flowered, publicly at least, under the
new. For a time it represented more than just a promising biological
substance: in the press and among its publicists in research and the
bureaucracy, interferon took on a political cast as a radical panacea that
had been ignored by a conservative cancer establishment. Interferon
indicated for the first time how potent the new, energized constituencies of biomedicine could be.

Interferon was the first, although hardly the last, of the biotech wonder products. "Don't get your hopes up, researchers say, but interferon
study is the hottest thing going on in cancer research," wrote the *Saturday Evening Post* in 1980. The magazine then made clear how foolish
it would be to listen to those persnickety researchers.

Could you be producing your own interferon in your crock pot at

27

home—as insurance against the day you might need it? Could a colony of bugs be manufacturing a life-saving interferon supply for you? Could you grow your own as easily as you culture herbs on your kitchen windowsill? It's a possibility worth exploring before cancer hits your home. One could conjecture that if you could treat yourself to supplemental interferon when you know you've been stressed, you might be able to ward off "the big C." The time to learn all about interferon is while you have the energy to do so—in other words, before "the big C" saps your strength and your strong will to live is all you have left.[1]

A rude shock awaited those who trusted the magazine rather than the researchers: *no one* knew much about the miraculous stuff in 1980. Interferon had been discovered twenty-three years earlier by a British scientist, Alick Isaacs, and his Swiss partner, Jean Lindemann. The pair were trying to figure out why animals rarely suffer from more than one viral disease at a time. They discovered that when one virus invades a cell, it stimulates the production of a protein that provides a short-lived immunity against other viruses. Discovery of that protein, which they dubbed "interferon," caused a minor scientific flurry and then gathered dust for the next decade or so, because producing even small amounts was prohibitively expensive. That, oddly enough, was a promising sign: if small amounts proved potent, *gallons* of interferon could yield untold wealth. Alas, the best that could be done was by a Finnish Red Cross team which began making interferon by infecting donated blood with viruses, then separating out the small amounts of interferon squeezed out by the white blood cells. The Finnish interferon was very dilute, from one-hundredth of one percent to one-tenth of one percent, and it required blood from some 90,000 donors to produce a gram, which then would sell for some $50 million. The Finns soon became the major world supplier. However, no academic lab could afford to buy even one entire gram—enough interferon to splash around in a thimble.

So for years interferon remained an object of interest only to a hardy band of perhaps a dozen virologists scattered around the world. This loose group operated far from the mainstream. Because interferon was so scarce, it could not be analyzed; and the material was so impure that all experimental observations were suspect. Interferon scientists developed a slightly fishy reputation for chasing an illusory panacea. "Isaacs was considered to be a serious virologist who'd become a crackpot, and dubbed his discovery 'imaginon,' " one scientist said.[2]

Then in the mid-1970s a subtle, but vital, shift took place. If viruses

Raising Consciousness

caused cancer—and one school of thought held that many, if not all, tumors were caused by viruses—and if interferon resisted viruses, then perhaps interferon could be used against cancer. That logic was promptly tested. An American scientist and interferon enthusiast, Dr. Ion Gresser at the Institut de Recherches Scientifiques sur le Cancer outside of Paris, tried it against cancers in mice, where a link with viruses had been established. It seemed to slow the growth of the disease. Unfortunately, other explanations seemed to account for the results, and the excitement died down. Gresser was soon followed by Dr. Hans Strander from the Karolinska Institute in Stockholm, who tried interferon on human cancer patients for the first time. He administered interferon to patients suffering from a rare, extremely malignant bone cancer called osteogenic sarcoma. Strander compared his patients' responses to historical results, or controls, which indicated that about 80 percent of those treated through surgery—which usually meant amputating a limb—would die within two years. Strander's interferon-aided patients seemed to beat those historical controls.

Strander's results were impressive, and the use of interferon against viral diseases took a back seat to cancer treatment. The fact that no one was really sure which viruses caused which human cancers was put aside; interferon was now taken up by the immunotherapists, who viewed it as a way of priming the immune system to destroy cancer—for enabling the body to heal itself. As a result, Strander's results soon became absorbed in the accelerating politics and economics of cancer.

Credit for the politics and promotion must fall, in part, upon two remarkable women. The first was a wealthy New Yorker named Mary Woodard Lasker. She came by her wealth through her second husband, Albert, who made a fortune in advertising before World War II. ("It takes salesmanship in print," he once said, "to weld every element of an advertisement—its ideas, its news, its drama—into a consummate whole and then to make it sing.")[3] The Laskers poured their wealth into medical philanthropy, supporting national health insurance and establishing the Lasker Awards for cancer research, a prize that, over forty-odd years has grown into a sort of first cut for the Nobel Prizes. In 1952, Albert died of cancer, leaving $11 million to his widow and $11 million to the foundation, which she took over. For the next thirty years, Mary Lasker operated out of offices across from the United Nations, cultivating presidents, legislators, and medical bureaucrats and convincing them of the benefits of spending for medical research. She was a major force behind the growth of the National Institutes of Health.

29

And she was active in the campaign that culminated in the War on Cancer. It was her brainchild, the Citizens Committee for the Conquest of Cancer, which, in a full-page ad in the *New York Times* in December 1969, first publically urged, "Mr. Nixon: You can Cure Cancer." Mr. Nixon, of course, eventually agreed.

In 1975, Mary Lasker was seventy-five. She still wielded enormous influence through a network in the biomedical bureaucracy known as Mary's Little Lambs. She was not an easy benefactress: she was a sort of George Steinbrenner of medical research, badgering NIH administrators to greater efforts, better results, and new directions, particularly as it became clear that her optimism was not being fulfilled. Curing cancer was not as straightforward as sending astronauts to the moon, and Mr. Nixon was not about to cure it by simply throwing money at it. Despite all efforts, cancer eluded understanding. Nonetheless, Lasker helped to nurture that special relationship between government and academic biology. And despite her age, Mary Lasker continued to work assiduously to keep it flourishing.

From that alliance came a Swiss-born geneticist named Mathilde Krim. Like Lasker, she was powerfully connected, particularly through her husband, Arthur Krim, a lawyer, former chairman of United Artists (which he and his partners had revived), a founder of Orion Pictures, and a power in the Democratic party. In the late 1960s, Mathilda Krim, who claimed that she had once run weapons to Menachem Begin's terrorist organization, the Irgun, was working at Memorial Sloan-Kettering in New York.[4] She was, by all accounts, an unremarkable researcher.[5] She proved, however, to be a master of cancer politics, a skillful promoter and shrewd manipulator of both the press and the bureaucracy. Like her husband, she had a gift for the deal and a sense of where power resided. Her first opportunity came in 1970, when she helped draft a report on the progress of cancer research for the Senate Labor and Public Welfare Committee, part of the coming wave of War on Cancer legislation. In preparing it, Krim ran across the interferon literature, an area, she came to believe, that had been undeservedly ignored by the powers at NCI.

For Krim, interferon fulfilled many different needs: personal ambition, politics, fund raising—and, of course, good intentions. Memorial Sloan-Kettering, for example, the oldest, largest, and wealthiest of the cancer centers, had been preeminent in cancer research in the fifties; in the sixties, it lost some of its power and prestige to the NCI. In the early seventies, however, a new approach to cancer treatment surfaced called immunotherapy. The discipline of immunology—the study of the body's immune system—had produced a series of fundamental breakthroughs over the previous decade; immunotherapy, then, was the attempt to turn those glimmerings of mechanism into application. Its most visible proponent was Dr. Robert Good, who had been appointed research director at the Sloan-Kettering Institute of Cancer Research (Memorial Hospital, the clinical wing, makes up the other half of Memorial Sloan-Kettering) in 1973. At the University of Minnesota, Good had pioneered various ways of stimulating the immune system to attack tumors, an approach dubbed "immunotherapy." With his arrival in New York, he and immunotherapy became famous. *Time* put him on the cover, and the *New York Times Magazine* published a long, admiring profile. "Today almost every puzzling disease in the medical handbook has become the target of the new immunological weapons," said the magazine.[6] Targets, alas, are not always hit. Nonetheless, Sloan-Kettering benefited from the publicity which brought in money, particularly since immunotherapy was not an area where NCI dominated. Besides, it was a therapy suited to the age: stimulate the immune system and kill cancer naturally without those nasty synthetic chemicals.

Those first immunotherapies—interferon had not yet made the shift to anticancer drug, and other immune boosters such as BCG and transfer factor were being touted—never showed much clinical efficacy. But the first immunotherapy boom clearly paved the way for interferon by showing how important the media could be in stimulating public support. Here was a way around peer review, that is, the practice of having other scientists review and judge scientific work and dole out money. By doing so, a new agenda-setting forum appears: the experts in the bureaucracy and academia were asked to share power with the press and the wider public. And, ironically enough, they were asked by scientists and bureaucrats.

It was now the mid-1970s. For over a decade, interferonologists had gathered sporadically to trade results. "Back then it was easy to know everyone," said Jan Vilcek,[7] a Czechoslovakian interferon researcher who settled at NYU Medical Center. In April 1975, Krim organized a conference on interferon at Rockefeller University. The size of the meeting, with about 200 participants, was not unusually large; the papers announced no breakthroughs; there was little scientific excitement. But participants were surprised and taken aback by Krim's elaborate arrangements and her promotion of interferon as a cancer cure. The press was invited. Krim worked to use a scientific meeting as a podium to the nonscientific world—as a media event. "When the word interferon came up, they didn't laugh anymore," she said later. "It was consciousness raising. Nothing else had occurred [scientifically]. There was not one shred of evidence in the literature [that interferon cured cancer]."[8] Ends justified means; people were dying; an obscurantist bureaucracy was bogging down the war effort. By November, NCI agreed to study interferon as an antitumor agent and buy a million dollars' worth or so for testing.

Krim herself organized an interferon laboratory at Memorial Sloan-Kettering. Meanwhile, the million-dollar NCI purchase proved inadequate. Dr. Jordon Gutterman of Houston's M. D. Anderson, an early activist in immunotherapy, began to lobby to get some interferon for his own studies. One of the first people he went to was Mary Lasker; convinced, her foundation bought a million dollars of it, and she began to lobby NCI and the American Cancer Society (ACS) to get behind interferon too. Gutterman also applied to the ACS, whose research chief, Frank Rauscher, had been appointed by Nixon to kick off the War on Cancer at NCI. At that time, the mid-1970s, Rauscher had resisted funding interferon research. Now he reconsidered. Gutterman was pressing for $2 million more, and Rauscher finally released the money, the largest grant the ACS had ever committed for a single project.

Krim took this as justification for her methods. "The American Cancer Society and some of the others felt they would really like to have something dramatic happen," she said later. "They felt they were making progress but it wasn't dramatic enough—not enough progress to satisfy the public. They saw chemotherapy and felt that that was not what they had expected to come out of all this. They wanted something that went whammo against cancer. I want something too. The reality is that we couldn't get it then or now, and we have to be very happy with what we have." To Krim, interferon represented as much marketing as science—a product to fulfill a public desire. "Everybody was very justified in wanting something whammo. I mean, some people think it's very chic not to want to treat cancer normally. Interferon isn't scientifically acceptable, but has some of the advantages of the popular nontreatments like Laetrile. It was exciting, mysterious."[9]

With the ACS purchase, this new coalition of administrators, fund raisers, and the press formed. Interferon appeared in the pages of *Time*, *Newsweek*, *Saturday Review*, *New York Times Magazine*, *Reader's Digest*, even in trendy *New York Magazine*, which declared confidently, "Interferon: The Cancer Drug We've Ignored." By the early 1980s, Krim had brought a new Hollywood constituency under her fund-raising tent.

Raising Consciousness

Laboratory workers recall phone calls from celebrities asking for interferon for relatives who suffered everything from cancer to herpes, and a Christmas party at the Krim apartment with Woody Allen and Anwar Sadat—heady stuff.

By then, what one observer calls "the interferon crusade"[10] had taken on a life all its own. Privately, among many scientists, doubts as to its efficacy lingered. As far back as 1975, the year of Krim's breakthrough meeting, questions had been raised about Strander's results. Bone cancers like osteogenic sarcoma are devilishly difficult to use in controlled experiments. Questions were raised about the selection of patients. And most devastating of all was the realization that Strander might have compared his interferon-aided patients against out-of-date statistics; in other words, his controls were faulty. True, osteogenic sarcoma *used* to kill off 80 percent or so of its victims in the two years following surgery. But by the time Strander went to work, procedures had improved enough to lower that figure, wiping out any benefit of his interferon therapy.

In the years ahead, interferon would prove not to be a singular phenomenon at all, but a family of proteins, a part, in turn, of the larger universe of interdependent and interactive proteins that make up the immune system. In the late 1970s scientists already categorized the interferons into three major classes—alpha, beta, and gamma—depending on the kind of cells that produced them. Each was different and each, in turn, could be subdivided further. Animal interferon did not always work in humans; human interferon did not work in animals. And each interferon seemed to produce different results, depending on the type or stage of cancer. Some interferons worked better with others, or with other immune proteins such as tumor necrosis factor or interleukin-2; others worked barely at all. All the interferons broke down quickly in the body, diminishing their effectiveness. As a result, they had to be infused in relatively high doses, which, in turn, created side effects: fever, nausea, lethargy. Its enthusiasts discounted these difficulties. After all, it was still far better than chemotherapeutics like methotrexate or adriamycin, which cause nausea, hair loss, and heart or liver damage. It was not a toxic chemical; it was a natural biological. "Interferon opens up a new form of cancer treatment which is nontoxic," Krim was still insisting as late as 1981. "There is no nausea, no vomiting, no diarrhea, no other side effects of chemotherapy."[11]

"The interferon crusade was successful because interferon was oversold," concluded a report by a Brookings Institution author in 1984.

"Certainly there was interferon hype, and all segments of the community participated—scientists who genuinely believed that they were on the right track and that money solicited at the expense of candor would be wisely used; investors and the public who wanted interferon to be a wonder drug and did not choose to ask whether the claims were overstated; and those representatives of the media who reported anecdotes with unbridled enthusiasm."[12]

Ironically, the money raised from NCI, the Lasker Foundation, and the ACS was soon overshadowed by the enormous sums raised on Wall Street. Krim was a transitional figure, clearing the ground for the new financing source, Wall Street. Companies by the dozen came to Wall Street seeking—and getting—financing, based on a vague promise to work on interferon; perhaps never before had more money been raised for a single, untested pharmaceutical. A few firms such as Genentech and Biogen worked hard to develop a recombinant DNA form of interferon; others tried and failed; others took the money and used it for other projects. Two major drug companies, the Swiss-based Hoffmann-LaRoche and Schering-Plough, spent millions on testing, dealing with the Food and Drug Administration, and building manufacturing facilities. And although interferon did prove that biotechnology could produce rare human proteins in large quantities, it has proven less like penicillin and more like Velcro. By 1986 when alpha interferon finally won FDA approval, the market for it was only a few million dollars a year.[13] Alpha was only approved for use against a rare form of cancer called hairy-cell leukemia, hardly justifying the investment dollars, although the picture would brighten a bit over time. Other applications have slowly followed, particularly against viruses—remember its days as an antiviral—and other members of the family await approval. But no one compares interferon to penicillin anymore.

As for the original interferonologists, center stage had its rewards and its irritations. "It is . . . an unrecognized law of science," Dr. Gerald Weissmann, the physician and essayist, once wrote, "that by the time its practitioners become 'ologists,' the field is already past its prime."[14] Such was the case here. The interferonologists found their work justified and rewarded. They were besieged by job offers, funding, awards, and consulting contracts. Reporters called them up. Television crews, lights and mikes bobbing above their heads, crowded into their labs and cramped offices. They were suddenly as famous as, well, as Mathilde Krim herself—or almost. A few plunged into ill-fated commercial efforts. Others took on consulting deals and stock options. But even as they did, the game was moving away from them. But what were they to do? "There was no evidence one way or the other," said Vilcek. "The only way to find out was to try it. The majority of those inside the interferon community were optimistic that something good was going to come out of this. Of course, you can't help but be pleased if people begin to pay attention to your field. There aren't too many people who are going to say, 'Wait a moment. This is going a little too far. Let's stop it.' That would be a little too much to expect."[15]

Interferon was a dress rehearsal for biotechnology. Krim sought to change the underlying rules by which the biomedical establishment functioned. She bypassed the conservative judiciary of peer review by appealing to the stormy power of public opinion. She mobilized the forces of the mass media. Interferon did not alter the basic relationship between academia, the drug industry, and the bureaucracy; but it applied the first major stresses. It created a space for entrepreneurs in academia and the bureaucracy, and it left a sense that a revolution—an insurrection against the prevailing conservative, bureaucratic powers—was not only necessary but possible. It also left in its wake the lesson that image meant more than substance, that means could be compromised to achieve desirable ends, and that a certain scrupulousness could be abandoned for expedience. All these were notions that dovetailed nicely with certain aspects of Wall Street. The forces set in motion by "crusaders" such as Mathilde Krim would bring together the tinder that would culminate in the Genentech offering and the subsequent explosion of biomania.

CHAPTER 4

Children of
Wall Street

FOR DAVID BLECH, as for so many others, the Genentech offering was a revelation. Blech was then a twenty-eight-year-old stockbroker, trained as a teacher; a part-time songwriter; an occasional investor in small medical stocks. He had not yet found his niche in the world. But on that day, he restlessly watched the ticker report the rise and fall of Genentech shares, and the possibilities leapt out at him. "I can do that," David told his older brother, Isaac.[1]

Isaac and David Blech were children of Wall Street. Born in Brooklyn, they had grown up among the jostling crowds that sweep across Wall Street; they knew, far better than recently minted M.B.A.s, the stereotypical basic emotions of fear and greed that drive Wall Streeters. Their father, Meyer Blech, had worked for many years at a smaller private brokerage firm called Muller & Company. Muller was a "retail" brokerage house, catering to a circle of private customers, occasionally sending a new issue into the world. Firms like Muller operate in a world of gossip and intrigue far from the thud of big stocks like General Motors or the whirl and crash of brokerage houses like Merrill Lynch or Salomon Brothers. At Muller, brokerage remained a very personal business: some stock trading, a deal, an investment. In 1980, before Genentech and biomania, David, a graduate of Columbia University Teachers College, was working at Muller selling stock and writing his

36

songs on the side. His brother Isaac, thirty-one, worked in advertising at a small manufacturing company in New York City; he wanted to make films. Together they dabbled in the market, David, thin-faced, lanky as a reed, with a shock of wavy black hair, resembled their mother, Esther; Isaac, who took after their father, was short, moon-faced, and bearded.

They had invested in stocks, including some biomedical issues, but that was just playing around. There had to be something more. Just before Genentech, David had picked up an issue of *The Sciences,* a magazine published by the New York Academy of Sciences. The issue, July/August 1980, contained an article by a Memorial Sloan-Kettering scientist named Lloyd Henry Schloen on the subject of monoclonal antibodies. Monoclonals, he said, had been all but passed over in the excitement generated by recombinant DNA. "It hasn't even been given the ultimate accolade of scientific journalism—'breakthrough,' " Schloen wrote. "Yet it is a technique that has scientists in private research buzzing. Advocates of hybridoma technology claim that the effect it will have upon medicine is comparable to the transformation of electronics by the transistor." What were its applications? Schloen discussed how, in the view of most immunotherapists, cancer cells displayed characteristic antigens, or targets, not shared by normal, healthy cells. Specific antibodies could target those specific cancer cells, thus fulfilling one long-elusive requirement for an effective anticancer drug: to kill tumor cells without harming normal tissue. Sloan-Kettering's Lloyd Old and his immunotherapy group, Schloen said, already claimed to have found such tumor-specific antigen in melanoma, a cancer of the skin. And Robert Nowinski, a former member of Old's team, now at Fred Hutchinson in Seattle, had actually reported curing leukemia in mice by pumping them full of antibodies.

Schloen had hit on several key words; he was offering a technology that was potentially as powerful as recombinant DNA, but that was still relatively obscure. He conjured up an image of a technology as sweeping as the transistor. He suggested that cancer might finally be cured simply and easily. The Blechs were intrigued and went out and hired Schloen to tell them more. He, in turn, led them to experts at Memorial Sloan-Kettering and, across First Avenue, at Rockefeller University.

The Blechs listened carefully. They came away with the sense that monoclonal antibodies might, as Schloen had suggested, have a grand future. The details of the science were not as important as the allure of the promise. The ability to generate monoclonal antibodies seemed to offer a way of making a "magic bullet," a long-sought-after substance first postulated by the great German chemist and microbe hunter Paul Ehrlich at the turn of the century. "The antibodies are magic bullets," Ehrlich had written, "which find their targets by themselves, so as to strike at the parasites as hard and the body cells as lightly as possible."[2] Genentech and Cetus and Biogen all had a big lead with interferons and human insulins and other esoteric, if powerful, proteins. But few companies had arisen to take advantage of monoclonals, of magic bullets. The name of the game was to find a scientist to lead the company they now actively began to plan.

As it grew and matured, molecular biology increasingly embodied a contradiction. It operated as if commerce and science occupied separate, antagonistic worlds, with academic freedom and the free interchange of data and ideas unable to survive in the harsh commercial world. The considerably murkier fact was that academics in scientific fields from physics to chemistry had long cultivated industrial ties, and even biologists themselves would sign up as consultants to the federal government, drug companies, or even to such controversial organizations as the Tobacco Institute. Where the line between academia and business was drawn increasingly became a matter of angry, often bitter, debate that overlapped the argument over biohazards. Moreover, these issues were exacerbated by the steady growth in the cost of doing science. Laboratory chiefs spent as much time raising funds as they did running their labs. As the 1970s drew on, power and fame increasingly meant the ability to mobilize large numbers of graduate students and postdocs—very cheap labor in a labor-intensive operation—to work on small parts of large questions. The process was circular: fame meant funding, which spawned more fame. Team leaders, laboratory directors, and Nobel Prize winners saw the demand for their services increase in direct proportion to their ability to attract money. It was not the same as running a business—profit played no part—but it was at least halfway there.

At thirty-six, Bob Nowinski, with his closely cropped beard and thinning hair, was an emerging figure in this class of entrepreneurial researchers. Born in Brooklyn, like the Blechs, Nowinski had a strong, aggressive personality: a quick wit, boundless confidence, and a real talent for articulating the romance of science. He had grown up as a self-described biological junkie, hanging around Memorial Sloan-Kettering, working summers as a technician in its labs. After a stint at Beloit College, in Wisconsin, he returned to New York and received a doctorate in immunotherapy from Cornell Medical College's Sloan-Kettering division. In the mid-1970s he moved to Fred Hutchinson—the "Hutch"—in Seattle. He struck people there as intense and cocky. Years later, the cancer center's founder, Dr. William Hutchinson (the cancer center was named for his brother, Fred Hutchinson, a former big-league ballplayer who succumbed to cancer), would tell of interviewing Nowinski and asking him his goals. "I want your job," Nowinski told him—an unusual, even startling, brashness in the scientific world of that time.

But he was good. As a postdoc, and then as a young professor of microbiology, Nowinski advanced to the front ranks of antibody research. He did not develop the first monoclonal antibodies, but he had the foresight to see their potential in cancer research. Capable of thinking in large, conceptual terms, he was an able fund raiser, quick to organize a team to follow his lead, to market the resulting research in papers and talks, and to set himself up as a scientific entrepreneur. He was, as Schloen's article indicated, one of the first to try to use monoclonal antibodies against cancer in mice; and he clearly saw the practical applications in humans. Sometime, somewhere, in those late 1970s he began to toy with the idea of business. It held a fascination for him; it provided an outlet for ambitions which, despite technical successes, were being bottled up in academia. Unlike most Wall Streeters, Nowinski knew of small companies, then still private, that were beginning to emerge. He began to talk about the possibilities inherent in a vehicle such as Genentech or Cetus, but focused on monoclonal antibodies.

He was not alone. In 1979, Nowinski received a call from Robert Johnston, a Princeton, New Jersey–based venture capitalist. Johnston had already formed a recombinant DNA company, Genex, in Rockville, Maryland, that resembled Genentech; now he wanted to assemble a monoclonal operation. He was in a hurry. On the West Coast, the venture-capital firm of Kleiner Perkins, the backers of Genentech, had already launched an antibody company called Hybritech. Nowinski had his own ideas about how such a company should be organized and about the strategic direction it should take. He found Johnston's offer attractive. Johnston, in turn, was pleased. He had already recruited a president and a director of marketing for the new company, which he called Cytogen; but Nowinski and his Seattle group were the hinge that would open this door. By October 1980, Nowinski was deep into meetings with Johnston and the management group in Princeton, while his researchers went house hunting. At the same time, the Wall Street investment house of Allen & Company tentatively signed on as an investor. Then, a snag: The deal went on hold while lawyers tried to straighten out a tax problem. The Seattle crew returned home.

It was November 1980. By this time, Isaac and David had emptied bank accounts, sold stock, and borrowed to put together $200,000. On November 13, a month after the Genentech public offering, they took out incorporation papers in Delaware. Isaac had come up with a name for this abstract entity: Genetic Systems. Their father then got on the phone and called Nowinski. Could his son David talk to him about a deal? Nowinski was interested and stopped in Manhattan before flying back to Seattle. They met, and David was so impressed that he quickly followed Nowinski back to Seattle—"Now that I'd met him, I wasn't going to let him get away," he said.[3] Beneath the differences, David Blech and Bob Nowinski had much in common—a feel for the deal, ambition, youth, and a sense of being outside but desperately wanting in.

In Seattle, David made his pitch. First, he said, you can run the company in Seattle; no need to settle in New Jersey. That pleased Nowinski, but still, the Blechs had a lot to prove—to him and to Wall Street. It was easy, after all, to dismiss these two deceptively casual and preternaturally young-looking men. Who were they? What had they ever done? Raising money—fast—was the way to dispel those doubts: the Blechs agreed that if they did not raise $1 million in six weeks, and $3 to $4 million in six months, the deal was off. To save the deal, Nowinski was reportedly given a check for several hundreds of thousands of dollars—which he could cash if it fell through. Just before Christmas, David sent off a formal offer to Nowinski: a salary of $100,000, a company car, athletic club membership, a free annual trip to Europe (not to exceed $5,000), and 1.2 million shares of stock. The deal was done. By New Year's Day 1981, Johnston discovered that he had been blindsided. Nowinski and his team would stay in Seattle. Without them, Cytogen collapsed.[4]

The Blechs had won the first round. How? They had moved with a sureness that belied their age and inexperience. Or perhaps they had won *because* of their age and inexperience. Compared to the competition, to Kleiner Perkins or Johnston or the investment bankers at Allen & Company, they were paupers. If Nowinski electrified any room he entered, the Blechs were shadows. They seemed deliberately obscure. What they had was drive, gall, a sort of offbeat charm, and, particularly with David, the strategist, a certain amount of vision. David recognized that the name of the game was packaging. Take the abstruse research-and-development work that Nowinski proposed and make it attractive to investors on Wall Street—people like themselves. The Blechs were less concerned with the patient construction of Genetic Systems than with getting to the market before biomania died and the money dried up. It had to happen fast. It was venture capital on the run.

CHAPTER 5

The Creation of
Pure Concept

THE BLECH BROTHERS were furiously busy between November 1980 and June 1981, when they sold stock to the public and laid the financial foundations of Genetic Systems. Compared to the speed with which the NCI would review a grant, the Blechs produced a considerable amount of cash in a short amount of time. Moreover, the kind of money they were accumulating made even the largest grants look small. And on the surface, it all seemed ridiculously easy. Bob Nowinski did not have to convince a jury of scientific peers that his approach made sense. He and the Blechs had only to convince investors, most of whom had never heard of monoclonal antibodies, that Genetic Systems had the key to a bright future.

The Blechs raised the money in stages, as if they were building a ziggurat or designing a booster rocket.[1] Mostly it involved the selling of stock and the convincing of investors to pay progressively more for a stake in the future. The classic problem in corporate finance is, How do you raise money without giving away excessive amounts of equity or future profits? How do you get investment capital as cheaply as possible? On Wall Street and in biotech circles, the financing of Genetic Systems would become a textbook case in how to solve this problem. It would, over the next five years, be studied and analyzed and imitated. Genentech was the great model operationally in the industry. But few companies had Genentech's assets; Genetic Systems, on the other hand, was a company that raised money with no hard assets, no products, no history to speak of. It was pure concept; it was the biotech everyman.

The process began before the Blechs even knew they had a deal, in the late fall of 1980; it climaxed in early summer 1981. Here is how it worked:

November 18, 1980: Preliminaries. In essence, the founders, in this case the Blechs, created an abstraction existing only on paper, and named it. The Blechs then simply made up, or registered, a number of shares, which they sold, at varying prices, to investors. The Blechs created 30 million shares within a shell they called Genetic Systems. (Most of those shares would remain in the vault, unsold.)[2] And five days after incorporation, more than a month before Bob Nowinski, the all-important scientific director, agreed to come aboard, Isaac and his mother, Esther Blech, each purchased 1,115,886 shares happily available at a penny a share. David bought a bit more penny stock, presumably for his initiatory role in the deal: 1,115,887 shares.[3] That hardly involved a lot of money—$11,000 each, an amount that could be raised with a couple of Visa cards. At the same time, new board members and consultants received anywhere from 100,000 to 10,000 shares each.

Not until January 7 did the most important Blech recruit receive his share. That was the date when Nowinski purchased 950,000 shares (his then wife bought another 200,000, and his parents and in-laws picked up a total of 50,000 more). From the shareholding perspective, the Blech family carefully insured their control of the company. They were, literally and figuratively, the founders of Genetic Systems.

February 3: Laying the foundation. The Blechs now looked further afield for a banker, some backers, and a front man. They discovered a banker in the J. Henry Schroder Corporation of New York, which bought almost 400,000 shares at about fifty cents a share—about $200,000; the price had gone up—plus an option for more. Schroder, a private New York investment bank, could pick up those optional shares at any time over the next five years for one dollar each. And Jeffrey Collinson, the chairman of Schroder, received a seat on the board. That same day, the brothers made $800,000 by selling 1.3 million shares at sixty cents each to thirty private investors, many of whom were New York City customers their father had long dealt with.

They had now raised a million dollars. In March they hired a president, James Glavin, a Harvard M.B.A., former pharmaceutical analyst, and veteran of several medical device companies. Wall Street knew and liked Glavin. On his arrival, he bought 200,000 shares at a penny apiece. At the end of April the Blechs sold off more shares to newly recruited directors; on May 7, the Blechs and their mother sold Muller & Company 100,000 shares, as the prospectus says, "in connection with the organization of the company"—payment presumably for that first introductory phone call and for those New York City investors.

May 19: The big score. The Blechs still had to make a big score to reach Nowinski's benchmarks: they had assembled a million in six weeks; now they had to raise $3 to $4 million over six months. In New York they had been talking to J. Morton Davis, the chairman of D. H. Blair & Company, about managing an initial public offering. As the underwriter, Davis would act as an intermediary, buying the stock from Genetic Systems and then, for a cut, assuming the risk of unloading it. Genetic Systems was now six months old. Simultaneously, John Simon, an Allen & Company partner who had been involved in the aborted Cytogen deal, approached Nowinski about a possible research venture with Syva, a division of Syntex.[4] Simon also told him that some Allen & Company partners might like to buy some stock. Allen & Company was a private investment partnership founded by a Wall Street legend, Charlie Allen. Syntex had begun as a seller of bulk steroids in the 1950s and ended up, under Charlie Allen's control, as a pioneer of the birth control pill in the 1960s—one of the great biomedical investments in history. Davis knew how much credibility a deal with Allen & Company and Syntex would provide on Wall Street. On the lookout for himself and his customers—the circle of investors who regularly bought what Blair was selling—he refused to agree to an offering until Genetic Systems had worked out a deal with Syntex and Allen & Company. Syntex and Allen & Company, on the other hand, had a big stake in a successful public offering. They wanted to wait until Davis and the Blechs worked out their arrangement for the offering before they agreed to buy the stock.[5]

Negotiations began. For four weeks they dragged on in three different venues: Nowinski with Syntex in Palo Alto talking about monoclonal antibodies; Glavin at Allen & Company in midtown Manhattan, negotiating the price of the stock sale; and the Blechs, stroking Davis on Wall Street. Nowinski and Glavin jetted from coast to coast. Contracts passed hands, lawyers met, and absurdly well-paid retainers cogitated over absurdly minor details. Tempers were lost and retrieved. "For all intents and purposes, Syntex and Allen & Company were one and the same. They were just in different parts of the country," recalled Glavin. "Meanwhile the pressure was building because we all knew the market wouldn't stay up forever. And Mortie wouldn't go ahead until we signed."[6] Paperwork in an offering takes weeks to complete, and those most attuned to the market, particularly the Blechs and Davis, could smell biomania giving way like experienced mountaineers sensing an incipient avalanche. Someone had to give.

Said Glavin: "It was late at night. We were having another meeting. We wanted to do the deal at five dollars. Davis said he would only do it at six—and only if Syntex and Allen had signed. We said that the Syntex negotiations were difficult, but he wouldn't budge. There was a lot of table pounding, particularly between the Blechs and Davis. Finally, Davis stood up and agreed to go ahead with the underwriting even though everything hadn't been straightened out yet. Davis showed a lot of guts." With the underwriting nailed down, the Syntex and Allen negotiations were soon wrapped up.

Thus evolved the deal that put the Blechs over the top: Genetic Systems sold 1.5 million shares to Syntex at one dollar a share. Allen & Company, as a corporation, picked up 300,000 shares, while various Allen partners, including Herbert Allen, the nephew of Allen & Company founder Charlie Allen, who was now running the firm, picked up more for their own accounts.[7] Here was instant credibility, Wall Street credibility. Genetic Systems *must* be the real thing if *those* guys are in, said observers.

The Allen deal also included a provision for stock warrants. Warrants are actually packages of stock and options to buy more stock at a fixed price by a certain future date. Like everything else, the price on the stock and the options were negotiated between Allen & Company and Genetic Systems.

June 4: The offering. The game now shifted to D. H. Blair. Mortie Davis had known the Blech family, through their father, for years. Blair also had a previous connection with biotechnology. A year earlier, it had pioneered the first biotechnology offering, a New York company called Enzo Biochem, which quietly preceded Genentech by several months. Blair, like Muller, operated in the over-the-counter market, catering to a retail clientele made up of mostly smaller investors trading for their own accounts. The firm had few dealings with institutional investors such as portfolio managers of pension plans or mutual funds, who tended to work with larger, more established brokerage houses, buying and selling more established companies. Davis, in particular, had built a reputation for taking large numbers of small companies public. When the Blechs looked for an underwriter, Mortie Davis loomed as the natural, even the inevitable, choice.

Davis was a bona fide Wall Street character. He had the sort of on-his-toes energy of a bantamweight, and he talked like some television version of a New Yorker. People would say: Who does Mortie remind you of? Maybe Tony Danza, but shorter. The little guy, the dispatcher in "Taxi"—Danny DeVito—but taller. Beneath the bombastics, however, Davis was very shrewd and very aggressive. He had clawed his way up from the bottom. Born Joseph Morton Davidowitz, he grew up in Brooklyn as the son of a kosher food distributor. In his teens and twenties he banged around as a fur stretcher, a diamond cutter, and a door-to-door vacuum cleaner salesman. At twenty-eight, he graduated from Brooklyn College. To pay for Harvard Business School, he waited on tables in the Catskills. In 1961 he joined Blair; in 1973 he exercised his options to buy the then-ailing retail brokerage house. At the time, it did not look like a very smart move. Wall Street was still staggering from the bust of the great bull market of the sixties, the go-go years. Small retail firms, everyone said, were dead.

So it seemed. The 1960s had been the biggest sustained boom on Wall Street since the 1920s. More and more of the country's growing wealth worked its way into stocks. Indeed, Wall Street brokerage firms grew so fast and so haphazardly that when the inevitable bust arrived, many firms simply went out of business, unable to handle the flood of orders. Those that remained had to spend heavily for computerized back-office systems. Many of the survivors, in turn, were forced to seek larger, wealthier partners. Wall Street adjusted and survived, of course, but in the process became a harder, faster, more grasping arena; the old-school clubbiness was passing. One of the institutions to emerge from the debris was an automated, reasonably well regulated over-the-counter market that fit somewhere below the more established New York and American exchanges and above the shadowy world of very cheap stocks, the so-called penny stocks. This new market was driven by the growing interest of investors, from individuals to traditionally more conservative institutional investors, in smaller, high-technology companies which, if they took off, provided the kind of blockbuster return—and the risk—that had now become *de rigueur* in an age of inflation.

That is where Davis took Blair. Davis recognized that the roles of venture capitalist and underwriter were increasingly blurred. When in the mood, the market would pay for any suitably hot concept. And so Davis collapsed the traditional incubation period most new companies underwent between incorporation and a public offering, allowing Blair and its investors to reap profits very quickly, while the company walked away with public financing. Blair specialized in the quick hit and unashamedly sold whatever the public wanted—"designer stocks" in the words of *Forbes* magazine, some promising, some dogs, although most had been around for so short a time or had promised such startling new departures that judging their prospects was like counting passengers on a speeding train.[8] Davis did well by it: *Fortune* magazine claimed he was worth $250 million in 1984; *Financial World* magazine estimated he made $60 to $65 million in 1986 alone.[9]

Blair's system depended on certain key ingredients: a crack team of brokers who could sell paint off a house; companies built around compelling, fashionable ideas; and regular customers who bought the stock when it came out and held it, willing to sit on paper profits. With the commercial prospects of many new issues still, at best, years off, keeping confidence aloft took on a high priority. Davis had to retain the illusion, and the illusion resided in the price. Davis often bought for the company account, but that alone could not keep every new offering aloft, even when there were not that many shares on the market. Admit that a stock was overpriced and watch it plummet like a bag of rocks. It resembled what William James called the will to believe: I will believe. I will believe. I *do* believe. One hot issue would lead to the next and the next. Everyone would profit. As long as you believed. So what if Blair could rarely engineer an offering the size of Genentech's $30 million, or the $120 million of Cetus. Why be greedy? He did a volume business. Smaller issues were more manageable, with fewer difficulties propping the price up.

These considerations determined the Genetic Systems offering. Through Blair, Genetic Systems sold one million *units* to the public at six dollars apiece. A unit, instructed the prospectus, consists of three shares of common stock and three so-called Class A warrants. One Class A warrant allows the holder to buy, for $3.25, an additional share of common stock plus one Class B warrant. One Class B warrant, in turn, can be redeemed any time before 1984 for another share of common stock at five dollars a share. What does all this mean? Strictly speaking, this could be called daisy-chain financing: one set of warrants triggers the next. Exercising all the options would produce nine shares of stock for about thirty dollars, a little over three dollars a share.

For investors it worked like this: With the shares he bought before the public offering, Herbert Allen also received 75,000 warrants. He could, at any time over the next five years, transform each warrant into another share of common stock for $1.50. If the stock was selling for just two dollars a share, Allen could make a quick fifty-cents-a-share profit by redeeming the warrant and selling the new share on the open market. Or he could wait for the stock to go higher. Warrants are thus an inducement to buy in early; it looks as if you are getting a lot, when actually you are receiving an invitation to participate in the future.

For Genetic Systems, this kind of warrant deal resembled a financial time-release capsule; the company would receive some cash up front, and then, at set intervals in the future several more infusions if holders redeem their warrants. Less beneficial is the dilution warrants generate; as warrants are redeemed, the base of common stock swells, diminishing the value of individual shares. Even worse if the stock tumbles, the warrants may languish, and the company may run out of cash.

All this is very clever, like a dress designed to flatter a rather spare figure. Not that there is anything particularly wrong with warrants; they are a very common, relatively straightforward financial instrument. The warrants not only offered a potential gain for investors but also gave Davis a stick to keep the company in line. As one Wall Streeter said, "By the time of the Genetic Systems deal, Mortie had learned to protect himself, to use warrants to insure a certain performance from companies he dealt with. By stepping up [increasing the price at which the warrants could be redeemed], he'd force them to do something with the money over the next few years." The warrants required Genetic Systems to remain attuned to Wall Street, to its fashions, its abrupt change of moods—to adjust to Wall Street's rhythms. Investors in common stock might want the kind of gains venture capitalists spent years working for, but they were rarely willing to wait as long for them. Venture capitalists could not easily unload their shares on bad news; investors in public companies could, and would.

The warrants also reflected the immature quality of Genetic Systems as a company. Certainly, Genentech and Cetus sold stock, and they worried over the stock market—they would have to return at some point for more money—but they were not tied quite as intimately as Genetic Systems. They had gone to the market when biomania was very hot; it was cooling now. And their perceived strengths on Wall Street obviated the necessity to play the warrant game. Both had been operating for a number of years, not months, and Swanson had actually squeezed out a paper profit from Genentech in 1979. Genetic Systems, on the other hand, had no earnings record, no history at all, unless one considers the eight months in which the Blechs wheeled and dealed. Its balance sheet before the offering had the simplicity of a Shaker chair: a million two in cash, $21,000 in other assets, $150 in deferred organization costs, and $1,500 in deferred offering expenses—no revenues, no profits, no products.

The prospectus tolled the risk factors: "the absence of an operating history; the lack of operating revenues; substantial dilution to public shareholders; and the fact that monoclonal antibody and recombinant-DNA technologies are relatively new fields." And, it all but admitted, Genetic Systems was going public long before it had any reasonable hope of products. "Although certain products have been developed in the laboratory by other entities using hybridoma technology, it will be a substantial period of time before such products can be clinically tested. . . . Even if the Company is successful in developing commercially saleable products . . . a substantial time may elapse before such activities can generate significant sales."[10]

The structure and pricing on the deal were determined by the rapid conception and birth of the company. The Wall Street professionals knew that Genetic Systems was a different beast from Genentech. "This thing was conceptually priced differently from Genentech," Simon from Allen & Company said later.

> Here, you had D. H. Blair doing the underwriting. They were at the opposite end of the spectrum [from Genentech's underwriters: Blythe Eastman Paine Webber and Hambrecht & Quist]. It was one of the first unit deals in biotechnology. As a device, they sometimes call units and warrants an opportunity for the public to invest in a venture capital situation. Well, that's one way to characterize it. It was clearly a highly speculative offering and was priced accordingly. The fact that it had Syntex helped its marketability, but no matter what, it was going to be priced low. It was a way of raising capital—and quite attractive to Nowinski. Genetic Systems set the pattern. Look, these guys [the Blechs, Nowinski] were terrific deal makers. Nowinski could sell the Brooklyn Bridge twice. But he also had good science, good people and some good contacts with a structure that would get diagnostics tested fast [through six Seattle hospitals].[11]

Biomania continued to roll, and the Genetic Systems offering on Thursday, June 4, 1981, turned out to be dramatically underpriced. For on that first day, the price of the units more than doubled to fourteen dollars. Once again, as in Genentech, someone had underestimated the demand. But Wall Street was getting cynical about the biotechs now. Instead of ascribing it to a wild market alone, many smiled and shrugged, That D. H. Blair—Mortie Davis knows how to hit that retail market, how to raise its temperature for an issue. In the months ahead, Genetic Systems would slowly sag—down to six in late September, as low as two a year or so later—in tandem with Cetus and Genentech. Indeed, Genetic Systems had just squeaked by, slipping under a slowly descending financing window.

The Blechs did not retire with their paper profits. They did not sell their stock and run. They did not forget Genetic Systems. Indeed, Nowinski and Isaac talked almost every night at eleven o'clock, New York time. But the brothers soon established their own spheres: David was the strategist, the thinker; Isaac, out of advertising, the marketer and promoter. And their attention soon strayed to other deals. In November 1981, two plant scientists left Campbells Soup in Camden, New Jersey, to form DNA Plant Technology, to apply genetic engineering to agriculture. The two largest shareholders were Isaac and David, with over a million shares each, at a penny a share, along with a few of their friends. By this time, the Blechs had become part of something called Schroder Venture Managers, Inc., a part of Collinson's Schroder. A few months later, in February 1982, they showed up again at newly founded Cambridge BioScience, Cambridge, Massachusetts. Once again, Isaac, David, and Schroder bought stock, this time at the price of *five ten-thousandths* of a cent. For 300,000 shares each, David and Isaac forked over $150. This was becoming a real science. The Blechs were no longer scratching for money. Even at six dollars a share, their stake in Genetic Systems was worth over $6 million each—100 times the $60,000 in original capital. And now with Schroder behind them, and a track record, the Blechs were becoming big players on Wall Street.

CHAPTER 6

The Community of Science

THE SUMMER OF 1961: The Kennedy Administration was in full bloom, dogs hurtled through space, a dispute over Soviet missiles in Cuba erupted. That summer, a year before winning the Nobel Prize, James Watson came down from Harvard, where he taught, to Cold Spring Harbor on Long Island's North Shore. In those years, Watson seemed to embody the expansive confidence of molecular biology. There was his groundbreaking work with Francis Crick, of course. But whereas Crick remained the chief theoretician of molecular biology, Watson—"Lucky Jim" as Sir Peter Medawar once called him—had a somewhat more public role, particularly in America. As he would later demonstrate, he could write not only seminal textbooks like his authoritative *Molecular Biology of the Gene* but also spicy, popular memoirs such as *The Double Helix*.[1] And he was gathering institutional power: he was already running Harvard's Biological Laboratories, and in a few years he would take over the laboratories at Cold Spring Harbor. Although he had all but ceased his own work, Watson already had the power to identify, from a storm of research, the key trends, the important questions. "I was just talking to him about the lab," the molecular biologist and 1975 Nobel Prize-winner David Baltimore once said, "and it occurred to me that it's a part of his genius—and it is genius—to be able to put together a laboratory to successfully carry out advanced biological research; to put

51

together the right people and the physical facilities, and to see the direction things would go, to understand the dynamics of intense, extremely bright people working together."[2]

Cold Spring, too, seemed to represent an ideal, at once expansive and exhilerating, of a community of science—the kind of community that many biologists would fear lost with the advance of commercial biotechnology. This ideal focused on free and unfettered communication: science was a communal endeavor, with a community of peers. Science, in fact, combined a brutal meritocracy with a theoretical democracy, and the Cold Spring complex contained both aspects. For much of the year it operated like any other full-time, competitive, academic laboratory. But in the summertime, the grounds, green and wooded and nestled about a tiny harbor and a bird sanctuary, were opened to the Cold Spring Harbor summer sessions, a kind of Wolf Trap or Tanglewood for biologists. The summer sessions had been going on since 1933—save for the war years—and they were dedicated to the kind of cross-fertilization that nurtured so much of academic science, particularly molecular biology. "Cold Spring Harbor in the summer grew to be a place where one could find people and be in prolonged useful contact with them," wrote Judson. "When Watson [first] went there it was informal, intimate, exclusive and even the play— swimming, canoeing, gathering clams or mussels, baseball in the evenings at the foot of the lawn, standing around talking on the road, beer at a place in the village called Neptune's—was saturated and preoccupied with science."[3]

In the summer of 1961, George Joseph Todaro, a second-year medical student at New York University Medical School, also came to Cold Spring Harbor. There was already, among the older scientists, a nostalgia for earlier days: the field was expanding, specialization was setting in with all its own barriers to communication, and the old intimacy was lost. But for Todaro, all was new and bright and exciting. This was an eventful summer for the twenty-four-year-old Todaro: Cold Spring, with its population of eminent scientists and bright students, would draw him in and set him on his career. "I was interested in research, but I was not yet convinced," Todaro said. "But there was one course that really seized my attention because we were allowed to work in the lab."[4]

The course was on tumor virology. Virology had a pedigree at Cold Spring. Watson himself had first arrived there in 1949 as a brilliant but, at nineteen, junior member of a loose group of researchers known as the "phage group," which was investigating newly discovered organisms called bacteriophages, or viruses that prey on bacteria. The phage group not only spurred many of the early breakthroughs in molecular biology, but it defined both a style borrowed from the glory days of quantum physics—collegial, intense, unstructured—and a methodology. They were preeminently model builders, a long way from the empirical drug industry. Faced with the numbing complexity of mammalian cells, like those in mice or humans, the phage group chose a simpler organism, the bacterium, to explore. And as a tool for that exploration, it chose one of the simplest of organisms, the virus.

Viruses were fascinating but elusive substances to study. Far tinier than bacteria, they were so simple in construction that they were consigned to a fuzzy borderline between the living and the nonliving. One could not call them dead exactly; they were certainly chemically active, but, unlike living organisms, they were incapable of self-replication. Early electron microscopes showed them to be tiny sheaths of protein with heads and tails, wrapped about an organic compound known as a nucleic acid. These phages, or viruses, were able to attach themselves tail first to bacteria, drill a hole through the membrane using an enzyme, then insert the nucleic acid—either RNA or DNA—within. In time, the cell would burst apart, or lyse, releasing hundreds of viruses exactly like the original. From external evidence alone, viruses seemed to be seizing control of the machinery of the host cell and using it to manufacture identical copies of themselves.

The exact role of viruses in cancer was a great mystery. Viruses are the Rosencrantz and Guildenstern of cancer research. They lurk in the shadows, now and then dancing forward, only to be yanked offstage once more. They were first linked to cancer in 1911—a decade or so after they were first discovered—when a young Rockefeller Institute scientist, Peyton Rous, thought he had linked a virus, now known as the Rous sarcoma virus, to a form of cancer in chickens. Few believed him, and a frustrated Rous later abandoned cancer research, calling it, "One of the last strongholds of metaphysics."[5] It would take fifty-five more years, until 1966, for Rous, then eighty-seven, to receive recognition for that discovery in the form of a Nobel Prize in medicine. In the intervening years, viruses rose and fell from favor. In the 1930s, a few scientists again became convinced that a link existed; after all, a few viruses clearly did cause cancer in animals such as mice and monkeys. Their optimism was summed up by British physician Dr. William Gye in a book optimistically titled *The Cause of Cancer*. If Gye was correct, then all researchers had to do was isolate the lethal virus and develop a vaccine to prevent it. Unfortunately, Gye was either wrong or simply ahead of his time—and it is a tribute to the problematic nature of cancer that scientists still debate the role of viruses in the larger human cancer picture.[6]

After Gye, interest in viruses receded a bit. Then, in the late 1930s came the work with phages. And by the late 1950s viruses were being used again as a way of penetrating the baffling complexity of the human cancer cell, and interest revived in them as a major potential cause of cancer.

Todaro's virology class worked to establish and quantify a link between viruses and tumors, a subject of increasing interest to Watson himself. Todaro and his classmates were dispatched to infect normal cells in petri dishes with viruses, then to try to determine how many of the cells turned cancerous. Such a procedure is called an assay, and it was a relatively sophisticated one for the day; not long before, the link between a carcinogen, say coal tar, and cancer was arrived at by smearing the stuff on a mouse and waiting for the lumps to appear. This, of course, was very crude and imprecise. Alas, this assay was working none too well. "The teacher of the course was named Harry Rubin, from Berkeley, and he was coming across the country," Todaro recalled. "The instructors kept saying, 'Wait until Harry gets here and he'll make it work.' Well, Harry did get there and basically what he did was to get the microscope focused right. And there, under our eyes, were transformed tumor colonies. I remember the moment, because I said, 'My God, this is a snap. Cancer has now become a simple problem because you can measure it. You know what the cells are, you know what the virus is, you know what the gene is that does it. Cancer wasn't such an impossible problem after all. This is what I want to do.'"

In 1939, literary critic Philip Rahv divided American writers into two camps: they were either subtle, refined, and precise "palefaces" like Henry James or Emily Dickinson, or "redskins" such as Walt Whitman or, in our own day, Norman Mailer: spontaneous, romantic, "half-baked mystics," wrote Rahv, "listening to inward voices and watching for signs and portents."[7] If George Todaro were a writer, he would have been born in a Rahvian wigwam, along with Watson and Crick, in symbolic opposition to the avatars of technique, the experimentalists—not a mystic particularly, half-baked or otherwise, but intuitive, a theorist rather than an experimentalist. Not that Todaro did not spend time in the lab, and not that a great experimentalist like Herbert Boyer could not spin theories. "The good scientist does both," explained Todaro. "It's more where your interests lie." Todaro's best work came not from

The Community of Science

shaking things up in test tubes, but from pondering the results. George Todaro—a very imaginative guy, echoed the scientific consensus.

Born in New York City, the son of a lawyer, Todaro went from the Fieldston School, a private academy up the subway in the Bronx neighborhood of Riverdale, to Swarthmore College, a small liberal arts school on Philadelphia's Main Line. From there he claims to have more or less wandered into NYU Medical School. Medicine did not exactly consume him. He was, for one thing, more interested in sports—captain of his high school football team, a good enough pitcher to get a tryout with the Yankees, and a member of the basketball team.

Today, Todaro, stocky and square, with olive skin, a shock of black hair, and soft dark eyes, carries the touch of the fog about him: soft-spoken, a bit hazy, he approaches a subject by working around it, jabbing it in a tentative way as if it might explode. He has a penchant for metaphors. A typical Todaro conversation goes something like this: "Yes, I see, well . . ." Long pause. "No. Let me see. It is as if the cell were, um, a telephone switching system." And the metaphor would unfold, aspect by aspect. He laughs easily, softly. At NCI, where competition and backbiting could be fierce, he became known for his interest in getting his name in print. Indeed, by 1983 he had racked up 200-odd scientific papers. Some saw his soft-spokenness as a cover for shrewdness—a bit too ambitious, it was implied, for his own good, although this complaint was hardly rare at NCI.

"I don't think George is the easiest person to deal with if you aren't certain what you want to do or what your capabilities are," said a former colleague. "He's not always the easiest guy to communicate with. But once you pick that up, at what level he moves, you can just have a wonderful interaction; he can be an absolute delight. He wasn't the greatest manager. His major strength is in good, productive scientific thought." Added another former colleague, "A lot of science has to do with human nature. How you're perceived. George is a very cool guy. Very reserved, but very shrewd. At NCI he used to disappear. I finally found out that he was slipping out to check the stock market. He knew all the symbols; he'd watch the tickertape. Down here [at NCI] that was pretty unusual. George is a funny guy. You know," he added, as if in explanation, "he's from *New York*."

Todaro's epiphany at Cold Spring convinced him to try research full-time. Looking back, his scientific career has a sort of logic, one step flowing into the next, which is really just the stuff of hagiography; science, like life, only looks simple in retrospect. Todaro's career began at New York University, where he worked with eminent molecular biologist Dr. Howard Greene. Todaro began infecting cells with tumor viruses, as he had at Cold Spring. Over the months, he developed more assay systems for other tumor viruses that further refined the process of quantifying the shift of a cell from its normal to cancerous state. That enabled researchers to cut down dramatically on the time and effort for experimental work, and to work with glass plates, instead of animals, focusing on individual cells, not unfathomably complex tissue systems. More significantly, he bred a cell line, developed from the connective tissue of mice, that grew extremely well in the cell culture systems used in those kinds of assays. This was important, for normal mammalian cells usually stop growing in cell culture, making assays difficult. These cells, on the other hand, were "immortalized," growing and dividing indefinitely—a trait they shared with cancer cells. Today, the line is known as NIH 3T3, and it is one of the most famous, and controversial, cell lines in cancer research.

Todaro gained a reputation with 3T3. While other young researchers were searching for projects, or jobs, Todaro had quickly hit upon a fertile area and moved on. The assays suggested a variety of possible directions involving viruses and tumors. And he spent very little time as a normal postdoc. "Things worked well," he recalled with a shrug and a chuckle. "They just . . ." the characteristic pause, "worked."

Back at school, in 1965, Todaro gave up the idea of actually practicing medicine. He was on a roll in more than just science. In his second year of medical school, around the time he went out to Cold Spring, he had learned about steroids, a family of naturally produced chemicals that perform a variety of regulatory functions in the body. And, around that time, he heard about a hot company out of Mexico City called Syntex which had done a considerable amount of work with steroids. Syntex had become famous for developing one of the first birth-control pills. "It was in Mexico," he recalled in a 1986 interview. "An over-the-counter stock. I bought it at twenty-six. It went up to a hundred and eighty. And I sold it too soon. I made a nice profit on it, but still . . ." He chuckled. "I have very fond feelings for Syntex. It helped me get through med school." Todaro finally received his M.D., then immediately entered the pathology department at NYU as an assistant professor. Two years later he was accepted at the National Institutes of Health. "It was a good time to be at NIH," he said.

NIH represented another vision of a community of science. If Cold Spring in the summertime was the ideal, NIH was the tough, competitive reality. Todaro entered NIH at the institute for allergy and infectious disease, the institutional logic being that viruses, whether they cause cancer or flu, were contagious. Soon, however, he shifted to NCI, a short stroll across the Bethesda, Maryland, campus. Todaro made rapid progress. In a year or so he was given a section, and a few years later, a laboratory of his own called the Viral Carcinogenesis Laboratory. Bureaucratically, he was now a figure to deal with. He had bodies under his direction, and the NCI system allowed him considerable freedom—and money—to follow his interests and instincts. Only results, in the form of important papers, mattered. Intelligence and hard work had taken him that far; but being the head of a lab requires other, less quantifiable, traits. The power in the field increasingly fell not to individual investigators but to laboratory chiefs. They were scientific managers, puppeteers, gathering a mix of young researchers, then gingerly positioning them. Much happened by serendipity, by a headstrong postdoc stubbornly wandering afield, or by a flash of insight; but the responsibility, and much of the fame, fell to the boss. A bit of fogginess and a loose rein were not necessarily deficiencies; overmanagement could suffocate a lab. The atmosphere was rarified; feelings, by necessity, were often bruised; competition crackled between postdocs in the same lab, between NCI labs, and between NCI and the rest of molecular biology. Success, for lab chiefs like Todaro, meant intuiting where the field was going: asking the right questions, suggesting the right line of experiments, hiring the right postdoc. A talent for theoretical work was an asset.

Todaro quickly showed an ability to synthesize masses of experimental data. In 1969 he began comparing notes with Dr. Robert Huebner, a physician and senior NIH scientist who for a time had been his boss at infectious diseases, where he ran the RNA tumor virus lab. They talked about a series of experiments Todaro had been doing. Todaro had been transforming 3T3 cells, turning normal cells into cancer cells. Then, an odd thing happened. Some of his 3T3 cells transformed *without* a virus present. And then, from some of those 3T3 cells viruses began to pop out. "I was perplexed," said Todaro. "I was sure we were messing up, that somehow viruses were getting into the plates. But then, the data got so convincing that there was no other explanation for it except that the cells themselves were making viruses; that the genetic information for making the virus was somehow already there." Huebner and Todaro talked round and round the problem.

Huebner was brilliant, but not in any conventional sense [said Todaro]. He was much more intuitive than rigorous. He could see three steps ahead and pull together five trains of thought. But he was not so good at proving each step; he didn't think that was as important. I'm that way, too—maybe more rigorous. I knew enough molecular biology to translate what he was saying into terms acceptable to biologists. I was the one who told him it wasn't enough to just say that viruses popped out. You had to speculate about enzymes, RNA, things like that. We had long, long conversations about philosophy and science. And out of that discussion, more than any experiments either of us did, came an idea.

The idea was a bit of redskin thinking called the oncogene theory, derived from *onkos*, the Greek term for "mass" or "cancer," and of course, *genes*: cancer genes. Todaro cannot remember who came up with the name. "It was not trivial," he said, "We discussed alternatives. Carcinogenes. No, that wouldn't work. Oncogenes. That worked." The theory offered this explanation for the phenomenon: Somehow there were certain genes in the three feet or so of DNA packed into a cell, that, sometime in the past, had been altered and left behind by invading viruses that had used the host DNA to replicate themselves. Cancer, or the sudden production of viruses, might then be caused by these genes suddenly switching on—or behaving in some way out of the norm. The oncogene theory not only made room for viruses in cancer and explained why it was so difficult to directly link them to the crime but also suggested the immensely powerful metaphor of the genetic switch for cancer hidden in the labyrinth of the cells. "Huebner's genius was this intuitive sense that if you push a button on, say, a toaster, the bread will pop up," recalled Todaro. "He couldn't explain the electronics, but the guys who could would probably never make the connection between the button and the toast. We did." The fact that this theory, this model, was not quite correct was almost beside the point. It was interesting; it explained the phenomena; it opened up a new path in this dense jungle.

Todaro and Huebner's oncogene theory was just that, a theory. No one, certainly not Huebner and Todaro, had ever actually stumbled across an oncogene in either viruses or human cells. But as a metaphor it was potent indeed; the beauty of a switch is that if one can turn it on, one should be able to turn it off. By then it was 1970. Todaro went back into the lab, trying to discover whether rat cells, pig cells, and other mouse cells would pop out viruses like 3T3 cells. They would, he discovered, but finding that was hardly as exciting as dreaming up the theory in the first place. He faced a dead end. "I probably beat it to death," Todaro admits today. "It probably wasn't the right direction to go. What I was doing was getting additional examples, instead of working on, say, the wiring."

Throughout the 1960s and into the 1970s, a small, elite group was forming at NCI: Edward Scolnick, Todaro, Phil Leder, and Michael Bishop, and a half-generation behind, Stuart Aaronson, Robert Gallo, and John Stephenson. They were all in their twenties or thirties; they were all, except Stephenson, physicians as well as researchers. Stephenson and Aaronsen had served as postdocs for Todaro before getting their own sections and labs. They competed; they collaborated. Some of them sat on the NCI coordinating committee that disbursed funds to other labs. They were brutally competitive. They were convinced that viruses were the way to open up the cancer mechanism, and they were increasingly obsessed with Todaro and Huebner's oncogenes. The age of the bacteriophage had passed—although not the concept of using a simple organism to pry open the mammalian cell. Now the virus of choice was a so-called RNA tumor virus or retrovirus, in particular, Peyton Rous's sarcoma virus. Retroviruses were unusual even for viruses. Instead of a protein coat enveloping a strand of DNA, a retrovirus contained a single strand of the messenger molecule, RNA. Some retroviruses were perfectly benign; others, Rous sarcoma virus among them, could transform cells to a cancerous state faster than any other known method. And, relatively speaking, they were simply put together: benign retroviruses contain three basic genes. The RNA tumor viruses, on the other hand, contained slightly more genetic material that was, quite naturally, implicated in the cancer-causing trait.

There was a central mystery to the retroviruses. How did they replicate without the genetic material, DNA? In 1964 a virologist at the University of Wisconsin, Howard Temin, had the temerity to suggest that the RNA in those viruses somehow ordered the DNA of the hostage cell to produce copies. Temin was talking heresy. Everybody *knew* that RNA could not order DNA around. That was a violation of Francis Crick's central dogma of molecular biology: DNA gives the order to RNA, which runs off to make proteins. With no mechanism to explain his contention, Temin's notion languished for six years. Then in 1970, Temin, and quite independently, David Baltimore of MIT, discovered an enzyme called reverse transcriptase that could take a strip of single-stranded RNA and fabricate a double strand of DNA that matched it, like a mill worker forming a symmetrical piece of metal from a template. That DNA could then order the cell's machinery to make viruses.

Reverse transcriptase provided a rush of optimism. Without it, a retrovirus would be unable to replicate and would be rendered harmless. Diagnosis of the onset of cancer, the thinking went, would require only the testing of a cell for reverse transcriptase activity. If it showed up, then cancer could not be far off. And a central role for reverse transcriptase suggested therapeutic possibilities. Halting cancer would simply involve blocking reverse transcriptase from acting—turning off the electricity that made the machine function. To that end, an Italian pharmaceutical company called Lepetit started talking up the possibilities of an antireverse transcriptase compound it owned the rights to called rifampicin.

All this was wonderful—but wrong. The biology was more complex than that. In a matter of months, a variety of labs, including Todaro's Viral Carcinogenesis Lab, found reverse transcriptase not only where tumor-causing retroviruses were present but also in cells infected by retroviruses that don't seem to have a cancer link. And, finally, they discovered it lurking in perfectly normal cells, particularly in cells from fetal tissue. Although different kinds of reverse transcriptase existed, the enzyme did not prove specific enough to be useful.

Enter the law of unintended consequences. So reverse transcriptase was not *the* key to cancer. But it was clearly another piece of the puzzle. And there was still the retroviruses link to cancer and to oncogenes that, Todaro and Huebner had speculated from a mass of circumstantial evidence, lurked within the genetic material. The trouble was that the human genome, the collection of genes in each cell, was fantastically, dauntingly complex: there were several million different genes in a genome, switching on and off like the largest telephone switchboard imaginable, ordering a myriad of protein products which then interacted in ways biochemists have not yet begun to figure out. Trying to find which of those genes were switched on in a cancer cell would be like riffling through an enormous pack of cards, looking for . . . what? No one knew what the cancer card looked like.

But what if you only had, say, ten cards to look at, instead of millions? And what if you were sure that one of them caused cancer? In other words, find the viral gene that caused cancer. That would be much simpler, and that is approximately what virologists Peter Duesberg and Charles Weissman set out to do. They chose the relatively simple and well understood virus that causes Rous sarcoma in chickens, looking for the trigger gene, the oncogene. By the early 1970s they found it and dubbed it the sarcoma oncogene, or *src*, pronounced "sark." Others followed with genes found in other viruses that caused cancer. One was called *myc,* because it caused mylecytomatosis in birds. Another was called *fes,* for feline sarcoma virus; it caused leukemia in cats.

These, of course, were viral genes that caused cancer in animals. Would those same genes also be found in human cells, circumstantial evidence that they had been left there long ago by wayward viruses (or, as now seems probable, that viruses had long ago accidentally picked up mutated cellular genes)? It is far easier to hunt for one in a million genes if you know what you're looking for, if you know what the joker in the pack looks like. Now that scientists knew what to look for, the evidence quickly appeared. Harold Varmus and Michael Bishop, both at UCSF, quickly announced in 1975 that the *src* oncogene was found not only in the Rous sarcoma virus but in chickens and, most significantly, in humans as well. Certainly, the evidence was circumstantial; but it did appear that genes in viruses that cause tumors *might* be related to genes in people that trigger a cell to become cancerous, a general restatement of Todaro and Huebner's original idea. Moreover, the fact that creatures as different as viruses, chickens, and people possessed analogous genes such as *src* suggested that they had appeared early on in the evolutionary drama, and that, when operating normally, they probably coded for fairly essential cellular proteins. By then, the oncogene field was in a ferment. Like a James Watson or a Cold Spring Harbor, this kind of science was peculiarly representative of the high church of molecular biology: ambitious, mechanistic, elegant in its simplicity, astoundingly complex in its details, and fertile in its implications. Not everyone agreed with the rapidly evolving oncogene model, of course, and clinical applications were still, at best, hazy, but oncogenes were well on their way to conquering the academy.

CHAPTER 7

Building Models

As MOLECULAR BIOLOGY matured, the competitive aspects sharpened. By the 1970s the frontier was moving forward quickly. Successful lab chiefs had to be acutely attuned to the scientific zeitgeist; spending too much time and effort on fascinating, but unimportant, initiatives could be devastating. Who knew what could trigger the downward spiral of a laboratory's reputation? A bit of bad luck, a few heroic, if losing, projects, and the lab could discover its stock in the academic community flagging. Soon it might begin to lose the best young talent needed to do the daily bench work. Each year it might begin to sink incrementally deeper into the backwater. Each year it would need, a bit more desperately, a major discovery to revive its fortunes. And each year that leap would grow more difficult. Soon the budget would be trimmed, grants would be turned down, the administrators would begin giving away space to some rising star, and everyone would be offering advice.

A lab chief has to continually refine some larger vision of the field, ruthlessly follow it, then sell the results to the greater scientific world. In 1976, George Todaro abandoned viruses. It was a calculated gamble, but he believed they had served their purpose; it was time to move on. He now turned his attention, and the efforts of his lab, to a group of newly discovered proteins called growth factors:

> I surprised my friends in virology by stopping my work with viruses. Here was a situation where some cell lines, including 3T3, released something into the medium that had the capability of stimulating

cells to grow. There were a couple of meetings of virologists at Cold Spring Harbor and I said that maybe the viruses aren't the way to go, but rather these proteins that cells make in response to the viruses, or coded by the viruses. They looked at me very strangely.[1]

Indeed, Todaro began thinking very hard about the cellular wiring, the metabolic circuitry of the cell, working his way back through the cell toward the oncogenes. In recent years there had been a general explosion of knowledge that threw light not so much on how the cell factory passes on information—much of that work was completed in the 1960s—but how it actually went about making molecules such as proteins, and how it uses those proteins to build, repair, and proliferate. The cancer phenomenon was central to that effort, because a typical malignant cancer cell—if such a typical cell exists—grows furiously and steadily, throwing all its resources into one grand and sinister goal: proliferation. Carcinogenic proliferation was, in this sense, just an extreme version of normal growth and development. Learning about the ways of cancer cells gave scientists insights into normal cells, much as Freud, rightly or wrongly, developed a theory of human psychology by studying hysterics. The unusual, the extreme, threw light upon the normal. The pitfalls—a tendency toward reductionism, a tendency to mistake effect for cause—were similar.

In particular, researchers focused on three classes of interrelated phenomena that seemed to be connected in the prevailing model of a "switched-on" cancer cell. First, of course, were the so-called oncogenes; in theory, normal genes that had somehow, through viruses, radiation, or chemical insult, been turned on. Second were the so-called receptors, whose role, like cranes on a loading dock, seems to be to seize specific proteins called growth factors bobbing past in the extracellular fluid, then pass them through the porous membrane into the thick, miasmic cytoplasm, the cell's factory floor that lies between the skinlike membrane and the nucleus, with its DNA. Third, and most mysterious and complex, was the chain of reaction that takes place within the cell when growth factor mates with receptor.

A few growth factors had long been known.[2] As long ago as 1924 researchers had recognized that insulin could stimulate cells growing in a culture dish. In the late 1940s a second factor was isolated from cancer cells that seemed to spur the growth of nerve cells. In 1975 research on a flood of new growth factors began. First came the so-called epidermal growth factors (EGFs), which were discovered while scientists were working with nerve growth factor. The EGFs proved to stimulate a wide variety of cells, cancerous and normal, that had large numbers of the growth factors' matching receptors studding their surfaces, like electric sockets awaiting plugs. EGFs triggered a variety of cellular responses but seemed to be most important in wound healing and cell growth. In fact, EGF receptors were found in the cells of animal species, such as fish, that go very far back in evolutionary history. That suggested that EGFs, like certain oncogenes, normally perform a very basic cellular function.

Cancer researchers were interested in any cellular event associated with growth. In normal cells, growth is stimulated when DNA orders the production of EGFs, which are released into the fluid surrounding the cells. Those cells with EGF receptors then begin "capturing" EGFs—the protein fits into the receptor like a key in a lock—and hauling them into the cell. That triggers a series of metabolic reactions, which in turn stimulates the DNA to make other proteins that further orchestrate cell division. Such activities can be as normal and natural as growth and healing. But researchers discovered that cultured cells that continue to receive EGFs begin to resemble cancer cells; that is, they continue to divide. Even odder, as Todaro's lab and others reported in 1975, some cells that have been transformed through viral or chemical means seem to lose the ability to bind to EGFs.

This is an interesting fact. Todaro and his group, particularly one of his researchers, J. E. DeLarco, began to feed EGFs to cells that had been transformed by tumor viruses. Soon the cells stopped absorbing the EGFs. Why? It turned out that another growth factor was actually competing with the EGFs to bind with the EGF receptors—a factor, presumably, that had been produced by the transformed cells themselves. When applied to normal cells, the mysterious protein allowed the cells to form colonies in soft agar—a gel-like material used as a culture medium—unsupported by any solid structure, another trait that separates cancer cells from normal cells. And then when the material was removed, the cells would return to their normal state. They called the material "transforming growth factor."

The resources at NCI came in handy for this type of work. Transforming growth factor (TGF) was released by cells in roughly the same way as interferon—a few molecules at a time. To harvest even tiny amounts meant growing lots of cells, then purifying and separating and measuring. Moreover, isolating the material was not enough. Todaro and his group labored to characterize the cellular pathway by which it worked: the receptors, the metabolic pathway, even the gene sequence of both protein and receptor. And, once all that was done, comparisons between other known factors and receptors could be made. The process required lots of time and lots of busy hands; fortunately, Todaro could command both. This was not intuitive dreaming; this was grind-it-out scientific work.

The results proved fascinating. Further research revealed that there were actually two TGFs, an alpha and a beta. The smaller alpha seemed to be the protein that was blocking the EGF receptor sites. Like EGFs, it had the ability to spur a cell to grow and divide. TGF-beta, on the other hand, seemed to allow cells to take on other traits of a cancer cell, particularly to grow in soft agar. Moreover, TGF-beta turned up in a variety of normal tissues, where it seems to have a basic function in wound healing. And, although TGF-beta does not bind to the EGF receptors, it seemed to have the ability to spur some cell varieties to create new EGF receptors; thus EGFs, TGF-alpha, and TGF-beta act synergistically.

By 1980, Todaro had a hypothesis to go with his TGF. He called it the "autocrine theory." Some cells, Todaro argued, particularly cancer cells, no longer require stimulation from the outside; they manage to manufacture their own growth factor, such as TGF, thus stimulating themselves to grow. Todaro called this an "autocrine," or self-secretion, mechanism. Cancer cells, in this scenario, engage in positive feedback, escaping the body's control and endlessly stimulating themselves to grow. Todaro speculated that TGF might have a normal function in a primitive, embryonic cell just a few days old. "Autocrine mechanisms for self stimulation would confer obvious selective growth advantages on early embryonic cells . . . when a critical mass of cells must be established rapidly," Todaro wrote in a 1980 paper.[3] Here, graphically, was the essential ambiguity of cancer that makes it so difficult to understand and to conquer: how inextricably bound this deadly process is with the forces of life.

The autocrine mechanism, when linked to hypothetical oncogenes, constituted a full-scale, if sketchy, model. It also suggested practical application.

> It is also obvious that autocrine mechanisms are potentially dangerous to the survival of the organism if they are not closely regulated as soon as they are no longer needed. . . . The recent isolation and characterization of defined polypeptide transforming growth factors, which appear to function by using such autocrine mechanisms, suggests that malignant transformation may be controlled some time in the future by means of specific inhibitors of the action of these peptides.[4]

In other words: As in reverse transcriptase, find an inhibitor, and you might be able to stop the cancer.

Still, models are, by their very nature, simple. Even as Todaro was writing, the evidence underlying the autocrine model was becoming, like so much else, more complex as researchers probed further. It turned out that other growth factors were also required to grow cells in soft agar, but these factors normally targeted cells that did not respond to TGF. Was this then a result of the artificial conditions of cell culture, or of hybrid cells like 3T3, or was there a deeper explanation? As the author of one major reference text concludes, "A caveat should be reemphasized [about the use of cell culture models]—although in vitro cell culture experiments can provide a significant amount of information about cell transformation mechanisms, they cannot tell us definitively how carcinogenesis works in the whole animal or patient."[5]

Nonetheless, the evidence of the TGFs was suggestive, particularly when it came to what Todaro called "specific inhibitors." Nature, Todaro believed, has a certain symmetrical logic, like clam shells, bird wings, eyes, nostrils, hands, and feet—like life and death. Nature, he felt—and this was far more intuitive than logical—has a propensity for balancing things out. If there is a protein ordered by the genes that jams the growth switch on, there should be one that turns it off, an antitumor growth factor. "Growth factors," said Todaro, "make the cells grow faster. That's not what we wanted. We wanted the opposite effect. To me it was very rational: You couldn't have a system that just sent positive signals. Nature is not going to do it that way. Again, as with oncogenes, it was obvious enough that you didn't need a lot of proof."

Again, Todaro made the imaginative leap. And again, Todaro's lab went hunting, this time for a protein that sent a signal to shut down growth proliferation. This was a far more difficult proposition. With TGF, the substance appeared first and was identified, before being fitted into a theory. In this case, the theory suggested that something might be out there. Nonetheless, in June 1982, Todaro's lab broke through again. Ken Iwata, who did the actual benchwork, and Todaro were listed as codiscoverers of a substance, the first of a family they called "tumor-inhibitory factors." Todaro would eventually dub them "oncostatins," or cancer stoppers. The name itself was charged with hope: cancer *stopper*.

The field was heating up. The ability to hunt through the tangled genetic haystack, to find a particular gene like the one that ordered a TGF or an oncostatin, and to produce it in quantity, was developing

66

rapidly. Biotechnology provided the molecular biologist with unheard-of tools for slicing open cells and looking around: transfections, a means of transferring DNA from one cell to another, and probes, the means to pull from the cell very specific DNA sequences. It created an enormous sense of power and possibility. On Wall Street, biomania was also heating up. In Washington, Ronald Reagan, dreaming of a return to American economic preeminence and productivity, was squeezing the budgets of federal bureaucracies all over town. The Reagan "revolution" had begun. Indeed, the notion of revolution was in the air. Hope sold stock; hope won elections.

By 1982, George Todaro should have been able to look back at his career with some pride. At forty-four, he had almost fifteen years at NCI; he had his name, either as a chief investigator or as the lab chief, on heavily cited, breakthrough papers such as "Oncogenes of RNA Tumor Viruses as Determinants of Cancer" or "Growth Factors from Murine-Sarcoma Virus-Transformed Cells" or "Autocrine Secretion of Peptide Growth Factors," all of which redounded to the glory of the institution, NCI, his lab, and himself. He had a lab; he had a reputation anchored in breakthroughs like oncogenes and autocrine factors; he was, in his own estimation, "a rising star" at NCI. His lab continued to forge ahead; the work with the TGFs and oncostatin was very impressive. He did not have to teach; he had not had to beg for money to buy a centrifuge or ten dozen white mice; he was able to work at top-notch facilities surrounded by the best talent available. Every few years, he confessed, he had desultorily cast about for a new job—"it's part of my personality" he acknowledged—but he had not found anything approaching what he already had at NCI. Ahead of him he should have been able to look forward to fame and awards and the respect of his peers.

But the game was so intensely competitive. Although he had arrived at NIH during the glory years, that aura was now fading a bit, through no fault of his own. Funding was not increasing at the same galloping rate it had in the mid-1970s in the early years of the War on Cancer. Indeed, the viral oncology program, which had grown into a major fiefdom within imperial NIH, was under fiscal attack. After receiving only $10 million in 1965, the year he had arrived, viral oncology had peaked at $110 million in 1979; that had fallen to $106 million in 1980, and $94.6 million in 1982. All this was made worse by inflation that was hanging around 10 percent. Indeed, NCI partisans argued, despite gross increases in budgets, the institution as a whole had not seen a true budget increase since 1975.

And there were other ominous signs. In January 1980 there was a changing of the NCI guard that anticipated the Republican tide that was about to sweep through the federal government. Dr. Arthur Upton, who had managed NCI in the late 1970s, was out; one of his staffers, the younger, smoother Dr. Vincent DeVita, was appointed in his place. Upton, while grudgingly praised for coping in difficult times, was criticized for not being more aggressive, for concentrating on abstruse, expensive molecular biology—not unlike the kind of work taking place in Todaro's Viral Carcinogenesis Laboratory—and minimizing less quantifiable areas such as nutrition. Indeed, viral oncology budgets were reduced because of a shift in funds to support research on the environmental causes of cancer. The prevention lobby, the anticigarette and nutrition folks, were gearing up. Upton paid the price for failing to achieve the ambitiously optimistic goals set by Mary Lasker and Nixon. DeVita faced a new era: although Reagan had not come out against the cancer program, it was an obvious example of a government program that had grown fat very quickly with few clinical results—few patients actually cured of anything—to show the folks back home.

All institutions react to a threat by producing defenses. Dictatorships become more dictatorial; bureaucracies become more bureaucratic. To survive in times of relative scarcity meant generating paperwork and memos, processing requests, coddling this office here, and nudging that one over there. One of the advantages of being at NCI—of being a scientific power at NCI—was that Todaro did not have to spend most of his day filling out grant proposals. He turned the spigot on in the morning, and the sink filled up with money. But now the atmosphere was changing: Need those mice? Fill out the mouse form. Need to hire staff? Work your way up the hierarchy like a politician on the hustings, then fill out forms—endless forms.

All good labs have a fearsome appetite for fresh minds and free hands. But now hiring was growing difficult. There were two problems; the first was internal. "If you wanted to fire or hire or promote someone," said Todaro, "you had to make up your mind: Did you want to administer for the next year, or do research? You couldn't do both. Everything took a year. You'd try to hire someone, and you'd never get a decision. You'd like somebody, but you couldn't make an offer. The inertia of the place was just overwhelming." There were telltale signs of future difficulties. "If you wanted a promotion, you had to demonstrate your worth on the outside at a comparable salary. Usually, you'd get an offer and bring it back and say, 'Look, *they* want me.' And you'd get what you wanted. But then, I started noticing that instead of taking the promotion, people started really leaving. As they left I thought: Maybe they know something I don't." The viral fraternity began to disperse. Phil Leder went to Harvard. Then Ed Scolnick left—to Merck, of all places. That was a shock, but many said Scolnick had made a big mistake to go to a drug company. To Todaro the loss of the sharpest of those fresh minds and free hands in his lab proved a very dangerous sign. "There was a time during the 1970s when you'd get lots and lots of applicants and you could pick and choose—maybe two or three positions for from fifty to a hundred applicants," he said. "And I just had one lab. That changed."

The second factor was external. It was one thing to lose talent to academia, or even Merck; it was another to lose applicants to companies like Genentech, Biogen, or Cetus. Todaro was a bit taken aback. He had, of course, heard of them. But he claimed to have paid very little attention to them, save as a minor irritant. Would they, he thought, be as secretive as the drug companies? Would they hide their results, refuse to present papers, slip in and out of meetings like spies? Would they try to feed off the open community of science and give nothing back in return? He shook his head. "I was slow to see what was happening. I think it was a function of being busy enough and not stepping back and seeing a major trend." Now, suddenly, these companies looked *competitive.*

Besides, no one got rich at NCI. His three children were approaching college age. Millions of dollars flowed through his lab, but he took home a civil service salary; the top salary for Todaro's class of tenured scientist in 1982 was $63,800.[6] Once again, Todaro began to look outside NCI. He talked to his wife and his children, who had grown up in the suburban quiet of Bethesda; they were willing to go. By early 1982 he began visiting universities and research centers. Nothing quite fit the bill. Either the facilities were not quite adequate, or the institution's research program did not quite fit his own. He did not want to teach; he did not want to sit on committees and write memos. He wanted compensation commensurate with his reputation. He had oncostatin, and he wanted to do something practical with it. He received an offer from M. D. Anderson, a well-respected, immensely well funded cancer research lab in Houston. Here was the greatest temptation yet—"A great offer, and very nice people," he recalled. He thought about it, talked to his family about it. He took to making lists: *Pro*—prestigious position, high-sounding title, good salary, fine facilities, no teaching, top-flight colleagues; *Con*—Houston could be so very hot. Todaro disliked hot weather. He turned it down.

Later in the year he heard from the Fred Hutchinson Cancer Research Center in Seattle. The center was seeking a research director. Todaro liked Seattle; it reminded him of New England. In the 1970s he had received an offer from the University of Washington Medical School but had turned it down; the action then was still at NCI. Still, the folks at Hutchinson were exemplary. He had even tried to hire a young guy working there several years back, Bob Nowinski. It had not worked out, although Nowinski had produced some antibodies against viral proteins generated by Todaro's laboratory. Todaro knew that Nowinski had, with some New York City backers, started his own company in Seattle. Perhaps he would give Nowinski a call when he flew out to see the people at Hutchinson. At the very least, Bob Nowinski was a very bright and entertaining guy.

CHAPTER 8

Fun with Numbers

WHILE GEORGE TODARO pondered his future, Genetic Systems roared ahead. Genetic Systems hoped to build its business on the fruits of early 1980s-style promiscuity: herpes, gonorrhea, and chlamydia. In 1982 the sexually transmitted diseases had attained an odd sort of notoriety. They remained a menace; the number of cases was increasing alarmingly, although in those relatively innocent days before AIDS, the sexually transmitted diseases no longer posed quite the dire threat to life and limb that they had in the past. But by developing into the house infection of the "me generation," the sexually transmitted—or "social"—diseases were simultaneously hyped and caricatured. If cancer was a tragedy ending in annihilation, the sexually transmitted diseases, particularly herpes, had about them a strain of dark comedy. One developed cancer inexplicably; one picked up herpes or gonorrhea from a blind date. Cancer triggered fears of an agonizing decline, herpes of acute physical and social discomfort. Cancer had the intellectual rigor of Susan Sontag and her *Illness as Metaphor*; herpes, the breathless sensationalism of the local news, Dear Abby, *Cosmopolitan*, and the Playboy Philosopher. They did share one thing: both were difficult to diagnose and treat. And both brought forth an almost daily diet of purported panaceas.

Genetic Systems' decision to pursue the sexually transmitted diseases placed it in a line of descent from Syntex. It was a sort of genealogy of sex: if Syntex and the Pill had ushered in the sexual revolution, Genetic Systems and its diagnostic tests would preserve the gains for swingers and singles. This was embodied in the deal that Genetic Sys-

71

tems forged with Syntex which paved the way for its initial public offering. Genetic Systems agreed to develop monoclonal antibodies for tests against five forms of the herpes virus—those that caused cold sores, chicken pox, shingles, mononucleosis, and genital lesions—as well as the microbial perpetrators of gonorrhea and chlamydia. To that end, Nowinski had won permission to use thirty-nine antibodies belonging to the Fred Hutchinson Cancer Research Center for a small fee, a rather standard licensing agreement.[1] At the time—several months before the offering—it was about the only tangible asset the new company possessed. Twenty-nine of those antibodies would, in turn, be used in the Syntex deal; the rest would be used in other projects. In return, Syva, the diagnostics subsidiary of Syntex, agreed to pay Genetic Systems $3.7 million over three years to defray the cost of research. And once the basic work was done, Genetic Systems would hand over the antibodies to Syva, which would build the actual test kits, organize the testing, pay the costs of regulatory approval, and then sell the kits. Then the arrangement would revert to a traditional licensing deal, in which Syva would pay Genetic Systems a percentage of sales.

Diagnostics run on volume. They operate off what one consultant calls the "ding principle" in medicine, the ding being the sound of an old-fashioned cash-register bell. "The more dings," this consultant would say, index finger raised, "the better." Imagine if every man, woman, and child in America were tested for cancer once a year. And imagine charging, say, two dollars a test. That would amount to over $400 million a year in sales, and figuring gross profit margins of 80 percent, not unusual in the business, over $300 million in gross profits. And that's not even talking about selling the test in Europe or Japan. And then there were the Chinese. There are a *billion* Chinese. And why limit this geographically? There are all kinds of cancers—ovarian, colon, lung, breast, brain; why not test for all? And there are lots of other conditions crying out for early detection . . . like sexual diseases.

Sexual diseases such as chlamydia seemed perfect for generating a multitude of dings, although not as many as cancer. As a 1983 report from the Seattle brokerage firm of Cable, Howse & Ragen said, "Diagnostics are useful only if the disease is prevalent, cannot be diagnosed efficiently with current methods and can be treated more effectively if diagnosis is rapid."[2] Was chlamydia prevalent? On the evidence of estimates that ranged from three to four million to ten million new cases annually, chlamydia gained a reputation of being the most widespread of all sexually transmitted diseases. Was it serious? At its worst, it could produce infertility in women or severe urinary tract infections in men.

Fun with Numbers

It was not, however, as one analyst declared, "often fatal," although about 11,000 females did suffer infertility annually from it.[3] How about treatment? Chlamydia could be knocked out by a ten-day regimen of tetracycline or erythromycin. The complicating factor, however, was that gonorrhea, a sexual disease more familiar to most physicians, resembled chlamydia quite closely. Gonorrhea, however, was vulnerable only to penicillin, which had little effect on chlamydia. Prescribe the wrong antibiotic because of a misdiagnosis, and the disease might linger on, quietly wreaking its damage.

Distinguishing between the two infections required a culture test, a relatively long, although quite accurate, process. It took about a week to grow microbes in dishes, or cell cultures, and apply the requisite drug. Then a trained technician had to determine which drug did the greatest damage—and make the diagnosis. Fortunately, but inconveniently, a week or so would not make all that much difference in the progression of a sexually transmitted disease.[4]

The argument made by Genetic Systems and its analysts for its sexually transmitted disease tests, on the surface, made perfect sense. Backing it up required statistics, figures, *numbers.* Wall Streeters, in particular, yearning after the certainty of mathematics, demand numbers. Forecasters thus raced to take the raw numbers on diseases like chlamydia; rake them together into round, impressive heaps; and extrapolate from there. This was standard procedure; everyone from the Congressional Office of Technology Assessment to the lowliest brokerage firm on Wall Street offered up their forecasts on various biotechnological products. In 1981, D. H. Blair published a report, "Genetic Systems Corporation and the New Age of Medicine," which discussed the big sales numbers that a rapid, easy-to-use sexual disease test could quickly produce. Blair failed to lay out its assumptions, which is really the only way to test the underpinning of forecasts, or cite sources beyond some figures pulled from *Newsweek* and *Business Week.* The report presumably subscribed to the belief that large enough numbers—any numbers—would provide comfort to investors. "Diagnostic kits currently represent $300 million in the United States alone," said the report. "In a recent article in *Business Week,* it was estimated that products based on monoclonals will carve a worldwide market in excess of $2 billion a year." Shifting gears, Blair also cited a *Newsweek* article predicting a $2 billion market for monoclonals used in cancer diagnosis.[5]

Impressive figures—but when would those big markets develop? And

73

how many competitors would leap in? And what kind of profit margins would those tests get? Blair avoided those questions and emphasized the positive. "We believe that Syva, one of the fastest-growing companies in the medical diagnostics field, will employ its strong marketing force to capture a substantial share of this market by being the first to introduce products developed by Genetic Systems," wrote Blair. "In fact, we expect Syva to make the first commercial sales of products developed by Genetic Systems in the second half of 1982, subject to prior FDA approval. Royalty revenues from this argreement will flow directly to Genetic Systems' bottom line."[6] Remember that date.

Blair avoided other issues as well. How much revenue might one expect to find piled there? How much real improvement would the Genetic Systems test provide over the culture method? How much marketing power would be required to allow it to bloom? Most importantly, how much would Genetic Systems be left with after Syva took its cut? Only close readers of the prospectus would discover the disturbing truth: Genetic Systems would get 5 percent of all sales, minus a small amount more which it had agreed to hand over to Fred Hutchinson for using its antibodies—something less than five cents on every dollar of product sold by Syva. For Genetic Systems, a 5 percent royalty meant it would reap $5 million in annual revenues *if* Syva sold $100 million worth of kits—a dim prospect. Syva's *total* business only amounted to $66 million in 1981 and only broke $100 million in 1983. If Syva was going to take off that quickly—more than doubling in three years or so—Blair should have touted Syntex, not Genetic Systems.

Why did Genetic Systems give it away? A variety of reasons existed: it needed marketing help, it had to have Syva to placate Davis and get the public offering, it believed its other products would have a bright future, and it was a little bit dumb. Nowinski, in fact, admitted the truth of some of this several years later:

> When we started the company we were heading for a public offering to raise 3 to 5 million dollars. Syntex came in with a research contract of 3 million dollars plus financing of a million and a half. It also was one of the principle mechanisms by which we were educated. We were a bunch of people from academia. Syntex provided a tremendous asset to us. We've been able to develop our disciplines—research, development, marketing—while working side by side with a very experienced group.[7]

None of this would have mattered much had a strategy not lurked beneath the deal. But there was a plan, and it was one that required, if not immediate success, at least a gesture in the direction of profitability. Unlike more mundane strategic plans—such as, I'm planning to sell these toothbrushes in those grocery stores for a dollar and make a dime on each one—Genetic Systems' strategy had a more complex, kinetic quality that, at the time, made eminent sense. The plan hinged on the ability of monoclonal antibodies to act as Ehrlich's magic bullet: to target antigens found on bacteria, viruses, or tumor cells, while ignoring others; to act both as a diagnostic, a means of identifying a specific cell or protein, and as a therapeutic, to destroy it. So the plan's logic stemmed not from commercial similarities between therapeutics and diagnostics—the two are distinctly different businesses with their products sold to different customers in different ways—but from perceived scientific synergies. It was a research plan forced into the attire of a business plan; it was putting the technology before the market. From the researchers' perspective, those two functions seemed all but identical. Finding a monoclonal that accurately identifies a particular kind of cancer cell means finding a way to deliver a deadly blow, to blast it out of there. The diagnostic would detect the problem; the therapeutic would kill it—all with the same antibody.

This notion had its genesis in the known structure of antibodies. Proteins might be thought of as the fuel that makes the cellular factory work. Cells undergo constant chemical activity, with proteins continually locking and unlocking, releasing and absorbing energy, cleaving in half, or providing the foundation for even more complex molecules. Beneath this complex interplay, however, lies an elegant geometry in which protein structure and function merge. For instance, a model of iron-rich hemoglobin looks like a somewhat lumpy beach ball with a sort of indentation at its center that attracts and fits a particular region of the oxygen molecule. Drifting within the electrical field of hemoglobin, an oxygen molecule finds itself sucked toward that indentation until it fits, hand in glove. The additional new molecule then causes the hemoglobin to twist, locking the oxygen in place.

Thus it is with antibodies and antigens. Structurally, antibodies are made of chains of amino acids which form proteins. Once formed, antibodies, like other proteins, fold up into three-dimensional shapes in space like molecular origami. All antibodies are shaped in a sort of rough Y, like a referee signaling touchdown. This referee, however, possesses four, not two, upthrust arms. Structurally, antibodies are made of four different components: two heavy chains that extend from the referee's toe to the tip of his upraised fingers, and two light chains—that second pair of arms. The referee's trunk—that is, from the shoulder down—tends to be alike in antibodies of the same class, hence its name, the constant region. What differentiates one antibody from the next? The exact shape of the cleft formed by the confluence of the light and heavy chains. This is the variable region, so called because its shape differs subtly from one antibody to the next, the reason that one particular antibody only fits its complementary antigen. Variable regions perform the same function as the indentation of the hemoglobin molecule: they are active sites, locks that hug only the correct antigenic key.

Monoclonals looked perfect for diagnostics. An antibody could, in theory, be dispatched to seek out a telltale clue to a disease—a particular antigen—and, like the baying of a bloodhound, indicate when it has found the antigen. Diagnostic tests fall into two categories: *in vitro*, those performed outside the body, or *in vivo*, those accomplished from within. They are very different. The blood test and the drug test are both *in vitro* tests, in which the evidence is removed from the body to be tested, like a fingerprint tested at a police lab. A barium enema, a brain scan, and an x-ray are considered *in vivo*: the machinery allows one to view, as if through a window, the drama within. *In vitro* tests can be simplicity itself. One takes a little blood or a little urine, mixes it with a prepared chemical, called a reagent, and checks the reaction. There are a variety of ways to tell if the antibody has found its telltale antigen in large numbers; it is enough to note that the patient need not be poked or prodded, bathed in x-rays, or injected with chemicals. No one ever died from an *in vitro* test, unless you count being scared to death from a false positive. As a result, regulation and commercialization of *in vitro* diagnostics move at a considerably faster pace than they do for *in vivo* tests. It takes about three months to gain approval for a new *in vitro* procedure, compared with two to five years for an *in vivo* one, which from the perspective of the regulatory authorities runs as many risks to patients as a therapeutic.

All this went into the formulation of Genetic Systems' master plan. The logic went something like this: First, we'll take our antibodies that show affinity for the chlamydia microbe or the herpes virus, or even various kinds of tumors, and develop them into *in vitro* tests that will allow doctors to detect these diseases sooner. We'll make our fast and easy money off that—and with our skills and technique, these should be tests the world will cry out for—and then we'll pour money into the development of therapeutics. In theory we could even employ the same monoclonals, chopping years off the development time. *In vitro* tests would help pay for therapeutics, from whence, of course, would come the big profits, and transform Genetic Systems into a fully integrated, smashingly profitable drug company. In short, the diagnostics business would allow Genetic Systems to leverage itself technically and financially into therapeutics.

Genetic Systems was neither the only nor the first company to try to leverage diagnostics into therapeutics. The first was Hybritech, a monoclonal company out of La Jolla, California. The moving force behind Hybritech was Brook Byers, thirty-three—the name rattling at the end of the recently expanded venture-capital partnership of Kleiner Perkins Caulfield & Byers. In 1978, Byers had been contacted by Ivor Royston, a professor at the University of California at San Diego, who argued the case for commercializing monoclonal antibodies. Byers in turn brought Kleiner Perkins with him. Eventually Byers hired an experienced executive, Howard "Ted" Greene, to run the place. In October 1981, a year after Genentech and four months after Genetic Systems, Greene took Hybritech public, raising a bit over $13 million.

Hybritech shared in the glory that was Genentech. Both had a reputation for going first class, of doing things right. Both combined an entrepreneurial flare with a businesslike aura. Greene knew how to push products out the door, and, like Swanson, he did not seem to be making things up as he went along. With Greene, Hybritech had one thing that few other biotech companies, besides Genentech, could boast: real professional management which ran the show—and with Kleiner Perkins, the comforting sense of a patron with deep and well-stocked pockets.

Coming at the end of biomania, Hybritech's public offering failed to make much of a splash. Ironically, Hybritech that year was a far more viable operation than Genetic Systems—or, heresy, even Genentech. It may have been the most viable biotechnology firm going, in that year. Byers was a venture-capital traditionalist: he adhered to the old-time gospel, in which venture money nurtures a young company by feeding it escalating doses of capital, culminating in a public offering, which funds the final stages of product development—that is, when you actually introduce a product. This was in contrast to the Blechs, who compressed the nurturing period to a few inconsequential months. The Blechs seemed to figure that they could worry about products later.

Thus, although Genetic Systems beat Hybritech to the stock market, it lagged years behind it operationally. Hybritech began to introduce diagnostic products soon after its public offering: an allergy test, a pregnancy test, some clinical lab tests. And Hybritech was hot on the cancer trail, which, at the very least, guaranteed good publicity. Greene was confident. In November 1982 he spoke to a group on Wall Street:

We believe Hybritech is part of a new generation of companies that are being created by basic changes in technologies and medical practices. No industry remains static in the face of major technological breakthroughs, and what's happening today parallels what Merck, Lilly and Pfizer achieved several decades ago when antibiotics were discovered. But even back then there were skeptics who said Merck would never make it. And today there are skeptics who say about Hybritech: "It's pie-in-the-sky" technology that's years away; "Hybritech's a research boutique," like most biotechnology companies; "Big companies will take over;" And "they're still a venture-capital deal" in the early stages of putting a company together. My objective is to destroy these misconceptions.[8]

In one important way, Genetic Systems and Hybritech were identical. "Hybritech intends to focus its development program toward products incorporating monoclonal antibodies," said its prospectus. "Longer term, the Company intends to shift this focus more towards new medical applications for antibodies as the Company's technology and revenue base grows and as the commercial implications of basic research with monoclonal antibodies becomes clear."[9] Although drearily phrased, it made its point: first diagnostics, then therapeutics; leverage.

Finally, consider for a moment that word *leverage*. On Wall Street, *leverage* is spoken of as if it were a newly discovered law of nature, profound and awesome, like financial antigravity. In classical mechanics it refers to the transformation of a small amount of force into a large amount of force, through the intervention of a tool, the lever. On Wall Street, money and force can be considered vaguely interchangeable, like Einsteinian matter and energy. Give me a couple of bucks, say the Archimedean financiers, and I will move the world—or at least buy the nice parts of it. Thus comes the purchase of large amounts of stock with a small amount of cash and a large amount of debt (one Wall Street definition of debt, the essence of financial leverage: OPM, Other People's Money). This approach may result in the takeover of large companies by small companies; or large takeovers engineered by insiders able to put up only a small amount of cash—the aptly named leveraged buy-out; or, as we have seen, venture capital—the transformation of a small investment into a large holding. In the evolution of companies like Hybritech and Genetic Systems was a fourth example of technological leverage: diagnostics would serve as the lever to propel—to *leverage*—Genetic Systems into pharmaceuticals.

Leverage, in short, was touted as the mechanism by which this revolution would be won.

CHAPTER 9

The Money Chase

SAY THIS about David and Isaac Blech: They assembled Genetic Systems with immense cleverness. Lacking an operating history, they sold a sophisticated package of credentials. The art of packaging is one of surface rather than depth, particularly when the depths are spooky and dark and full of the unexpected.

For Wall Street, the Blechs built a facade studded with references to other praiseworthy examples of science and business. First, they picked a technology—monoclonal antibodies—with great promise and few companies dedicated solely to its commercialization. Then there was Dr. Bob Nowinski, the scientific director with the expertise in monoclonal antibodies. Nowinski soon went on the investment-meeting circuit, meeting brokers, investors, and analysts, trying to build enthusiasm for Genetic Systems. He proved to be a marvel, placing the new firm into the long sweep of history by giving a spellbinding little talk on six revolutions in science: chemistry, physics, psychology, medicine, electronics, and now, biotechnology. "Nowinski," said one analyst, "was the best explainer of science I ever heard." There were no long, abstruse explanations required. All one had to do was mention magic bullets and show the numbers, particularly to the analysts. Those numbers—on herpes, gonorrhea, and the more obscure chlamydia—seemed to indicate that some large, untouched markets existed out there. And beyond that waited the promised land, cancer.

Finally, there was the scientific advisory board. Such boards were endemic in biotechnology. A Nobel Prize, preferably in medicine or

80

chemistry, was the ultimate credential for a board member. Like so much else, the roots of the advisory phenomenon were planted deep in the soil of academia, which operated on the basis that advice would be freely given and accepted. True, the drug companies had long used academic consultants, but their share prices hardly fluttered when they announced hiring a researcher like, say, Edward Scolnick. The stars of the drug world were businesspeople, not consulting scientists. Perhaps Mathilda Krim's use of a scientific meeting to raise consciousness foreshadowed the change. Whatever caused it, in an industry that preached the ascendancy of science over business—and the pure Silicon Valley-style entrepreneur is in many ways antibusiness, as the romantic is anti-intellectual—scientific boards carried great weight.

Genetic Systems snared no Nobels, but its scientific advisory board had clout. The names might have been strange to Wall Streeters, but their institutions had the ring of authority: Matthew Scharf, Albert Einstein College of Medicine; Robert Good, formerly of Memorial Sloan-Kettering, now at Cornell University Medical College, Rockefeller University, and the University of Oklahoma; Norton Zinder, Rockefeller University; Dr. Avrion Mitchison, University College, London; Dr. King Holmes, U.S. Public Health Service, Seattle.[1] If one took these names at face value—as Blair did in its Genetics Systems report when it declared that this illustrious board would actually guide company research—one could not help but be impressed.

Actually, there was less here than met the eye. Not that the reputations were not genuine: Holmes was an eminent expert on infectious diseases; Scharf and Zinder, world-class molecular biologists; Good, the pioneer of immunotherapy. Rather, like most ad hoc committees, this one possessed less heft than the sum of its weighty parts. The board gathered intermittently to review progress, now and again tossing a bright idea in, or offering a bit of wisdom. And, like more traditional consultants, they were available to discuss particular difficulties. But with the exception of Holmes, whose laboratory and pool of patients were nearby in Seattle and whose specialty, infectious diseases, was an early priority at Genetic Systems, the real research was forged by scientists in their twenties or thirties, directed by Nowinski. In fact, the true role of the scientific board was less to advise and more to legitimize the scientific credentials of the company.

The surface mattered to Nowinski, too. Particularly in those early years he cut an articulate and impassioned figure. Many who met him came away impressed, at the very least, by his personal magnetism. He had an infectious enthusiasm for the endeavor and for the medical possibilities. He seemed, in the early days, to be everywhere: upbeat, energetic, idealistic. We are all in this grand adventure together, he seemed to say. He learned quickly, soaking up what seemed to be reams of new data, continually spinning out new ideas. And if he was headstrong, that came across as just another side of his enthusiasm. He talked about building a team, and he successfully cultivated a sense of participation in a grand enterprise.

Biotechnology was new; there were few companies like Genetic Systems. The future was exciting to contemplate; the present, loose, exhilarating, and creative. No moldy, old corporate fustiness here. Not long after setting up shop, Nowinski reinstituted the common university practice of Friday happy hours, like Genentech's Ho-Ho's, once a month. His practice of stopping at a local doughnut shop called Winchell's evolved into a weekly ritual—every Friday at ten—called Treats: coffee, doughnuts, and orange juice. Over time, full-fledged retreats were held twice a year. He preached the gospel of the entrepreneurial, science-based organization. He had what associates describe as vision: he talked about how Genetic Systems could take a radical, a revolutionary, technology to create new medical advances; about how this would be medically important, commercially successful, and financially rewarding; about how a company could heal and enrich lives.

He cultivated individuality. He dressed as if he were consciously mixing the sartorial symbols of science and business: suits and Reeboks, padded shoulders, pleated pants, cable-knit sweaters, and black leather jackets. "He can be a captivating guy," said a Wall Street analyst who met up with him in those years. "The quintessential scientist. A great marketer of himself. Scientists who develop a presence are impressive. They walk in the room, they seize everyone's attention, tell a great story, get everybody pumped up. Nowinski was like that. Every time I would see him he would have gone a little more bohemian. He never wears coats. He shows up in his classic *black* outfit. He's a real classic—and extremely bright." He filled his office, facing Puget Sound, with plants and a collection of clocks. He had a stereo system. He went first class. Every two weeks he had a new flower arrangement brought for his office. Events were catered by Gretchen's Of Course, a trendy Seattle caterer. He drove a company car, a Ferarri. The building, planted among waterfront warehouses, was modern and new; rounded, with smoked glass, like a hockey puck with sunglasses. Up the hill loomed Seattle's Space Needle from the 1962 World's Fair. To the west, a stone's throw, was Puget Sound. It did not in any way look like a garage. This was a company for a new age. Appearances mattered.

The Money Chase

As 1982 rolled in, Genetic Systems faced a problem common to rural banks, Third World countries, and the federal government: a lot of money was pouring out the door, while only a little was coming in. The company was operating on a form of deficit financing. The lease on the four-story waterfront building amounted to $277,000 a year.[2] Nowinski and Glavin had been hiring furiously; by the end of 1981, Genetic Systems had thirty-six employees, twelve with Ph.D.s or M.D.s. Labs had to be built and equipped with petri dishes, pipettes, centrifuges, white coats with names stitched over the pocket, and multitudes of mice. The UPS courier downstairs with a truck full of gel tanks, cell counters, deionizers, fraction collectors, slab dryers, and microscope attachments had to be paid.[3] And even as the first projects began, Nowinski and his team were planning product lines to go with sexually transmitted disease tests—ones for respiratory infections and opportunistic infections.

In the language of the venture capitalist, the company was "burning" cash. To get an idea of the situation, it is useful to refer to an accounting concept borrowed from the venture-capital community called the burn rate. The burn rate is a benchmark developed for companies so immature, and so far from the marketplace in terms of products, that revenues and earnings, which normally account for a firm's financial condition, have no meaning. Venture capitalists might chatter on about projections of sales and earnings over lunch, but at night they figure burn rates.

The burn rate, and its various offspring, provides a snapshot of how fast a company is spending its capital—burning it up, like trash in an incinerator—on research and development. By getting an idea of the overall burn rate, one can then subtract it from incoming dollars from research contracts, the sale of stock, or venture-capital infusions. The resulting figure, usually a negative one, is called the net burn rate; the larger the negative, the more serious the capital erosion. That, in turn, leads to a more chilling measurement, the survival rate, which tells one how long an operation can remain viable with its current rate of spending and financial resources before seeking protection in Chapter 11 bankruptcy.[4]

Genetic Systems turned the heat on high. Genetic Systems was technically a development-stage company, which essentially meant that it was expected to burn money in order to build for the future. All biotech companies fit this description. Indeed, in years to come, biotechnology analysts would try to value companies as stocks by adding up the research dollars spent—the theory being that the more you spend, the better you must be. Not everyone agreed with that reasoning, which failed to take into account any measurement of effectiveness or productivity. "That's like suggesting that you can increase your net worth by running up the balances on your credit cards," said one observer. In the last six months of 1981, Genetic Systems had spent about $1.4 million and had about $6 million in the bank. Add to that Syva's $277,000 payment for the sexually transmitted disease tests (just enough to pay the rent), and Genetic Systems was burning cash roughly at the rate of $240,000 a month through its first year.[5] At that rate, Genetic Systems had enough cash for a little less than three years of solvency—low for a diagnostics company, which with its lower research, testing, and regulatory costs tends to burn up money more moderately than firms developing therapeutic drugs. It was, however, about average for more free-spending therapeutic operations such as Genentech or Cetus.

Thus, almost as soon as the public offering was completed, Genetic Systems went chasing money again. The responsibility for this now swung to Nowinski and his president, Jim Glavin. The Blechs, although available, were back in New York chasing new deals. There were two financial roads to travel: debt, where investors such as banks or bondholders lend money, or equity, where investors actually buy a piece of the action. Bankers, unless they work out of shopping malls and drink champagne from their boots, look at startup companies borrowing large amounts of money as, at best, reckless, at worst, deranged. Debt also imposes another future bill, in the form of interest, to pay. Likewise, bond issues of startup companies, even in an age of junk bonds, do not exactly wow investors. For Genetic Systems that left equity. In 1982, this option did not look very feasible either. The Reagan recession had flattened the market, and biomania, based as it was on expectations still beyond the time horizon of most Wall Streeters, had now fully died down. Fortunately, the machine the Blech's built—on a design by Mortie Davis—proved remarkably effective. Those Class A warrants, due in June, could be turned into common stock at $3.25 a share; by the fourth quarter of the year the company had received a bit over $4 million as payment for 1.3 million Class A warrants, pumping up the company's cash position to almost $10 million.[6] At the same time, Syva added another $1.2 million, or some $100,000 a month, on the sexually transmitted disease project.[7]

That was a start. Next the company used another vehicle that fell somewhere between debt or equity: the limited partnership. At the end of 1981, Genetic Systems formed a subsidiary called Respiratory Diagnostics, Inc., which, in turn, became the general partner of a research-and-development limited partnership called Genetic Systems Respiratory Partners, shares in which were then sold to a small number of investors.[8] This was slick corporate finance on the order of the warrants, and it was served as a way to wriggle free from the grasp of the stock market. The thirty-six partners paid $3.4 million to fund a number of specific diagnostic projects—monoclonal-based tests for Legionnaires' disease, strep throat, three forms of pneumonia, and three varieties of the common cold. Again, the litany of numbers appeared. "Respiratory infections," said a later prospectus, "are the most prevalent form of infectious disease in the United States, resulting in one hundred million office visits or hospital stays a year, and the third largest cause of death after heart disease and cancer."[9]

The limited partners had more power than an average shareholder. The partnership actually owned the products that resulted, agreeing to license them back to the company for royalties on sales which would then be disbursed. The investors also received a snug tax shelter; in theory they would take the losses incurred in the research years as a personal tax deduction, then collect when the products made money. More likely, they would be bought out by the company; that is, given stock in exchange for their partnership shares. What were the advantages to the company? In effect, the government, through the tax deductions, would be helping to finance its research. And Genetic Systems could receive money for research without giving away huge chunks of future products, diluting the stock (already recently loaded down with the Class A warrants), or depressing the price, an event to be avoided with those unredeemed B warrants floating around. Such partnerships thus resided "off the books," having no effect for most of their duration on the company's profit-and-loss statement.

Partnerships could be a risky proposition for investors. The potential for fraud always existed. Nonetheless, they were quite common in the oil and gas industry and among Silicon Valley electronics companies. If a project was successful, a partnership would eventually pay off through royalty payments, or by trading partnership shares for common stock or cash. In each case, either the value of the common stock or future earnings were watered down—in financial terms, diluted. In particular, the dilution of shareholders' stock—the reduction in value of individual shares that takes place when the total number increases—could be immense and invisible, like some dark planet detectable only by the tug of its gravitational field. Moreover, such partnerships tended to overlook potential conflicts of interest. Company executives actually managed the partnership as the general partner, while perhaps investing as individual limited partners. When it came time to buy out the partnership, whose interest would management represent—the limited partners, themselves, or the common shareholder?

Genetic Systems was one of the first biotech companies to attempt an R&D limited partnership, albeit a relatively small one. The big splash would come that same year when Genentech sold a $49 million partnership. Genentech's Swanson argued that money raised from a stock offering should not be used to fund product development; rather, it should be preserved, or at least used in the final stage of commercialization, when profits beckoned. To use that money too soon was to expose oneself to the vagaries of the market. Thus, by raising money through the partnership, Swanson could, in theory, protect his publicly raised nest egg. This was a lesson other companies, including Genetic Systems, failed to heed. Money was money, to be spent wherever it came from, and to be applied as early in the development process as necessary. After all, if it ran out, there would always be more where that came from.

Swanson, on the other hand, knew the free-spending attitude on Wall Street would not last forever. He stockpiled cash and applied partnership money to specific products just entering clinical trials, when the risk of failure had been reduced. The purpose of the first partnership—Genentech would eventually float four of them, raising $126 million—was to finance clinical testing of human growth hormone and gamma interferon; Genetic Systems, on the other hand, was raising money to do basic lab work, with the hurdles of testing and regulatory approval looming ahead.

Compared to Genentech, Genetic Systems' first limited partnership was strictly minor league. But the cash came in handy. Meanwhile, Glavin and Nowinski were out wooing new corporate partners. These partners would pay Genetic Systems to perform certain research tasks, then pay a royalty on the resulting product. In March, Genetic Systems agreed to develop monoclonal antibodies for Cutter Laboratories, a subsidiary of the West German drug and chemical company Bayer AG. The monoclonals were aimed at gram negative bacteria which cause infections that are resistant to hospital antibiotics, often threatening immuno-suppressed patients. The deal had an interesting spin: Cutter agreed to pay $2.4 million to Genetic Systems over three years, then royalties on any therapeutic product sales. But Genetic Systems could keep the antibodies for any *diagnostic* kits it might want to market.

The Money Chase

More deals were in the works. In August, Daiichi Pure Chemicals, a Japanese chemical company, paid $125,000 for rights to market in Asia a diagnostic kit measuring various constituents of the immune system. A month later, it agreed to market products Genetic Systems was developing to diagnose immune system problems. Two months later, New England Nuclear, a subsidiary of DuPont, picked up the marketing rights for those diagnostic products in the rest of the world.

By the end of the year Genetic Systems had over $11 million in the bank, a variety of corporate partners, and a web of related research contracts. And it was spending money at the rate of $328,000 a month— that is, the burn rate was $328,000 a month.[10] That gave the company a bit more than three years to live—a slight improvement over a year earlier. With the Cutter deal, the company might even begin to recoup from the unfortunate terms of the Syva sexually transmitted disease test deal, with its meager royalty rate; and the limited partnership won some freedom from the stock market.

Still, it took its toll. If research burned money, fund raising burned time. Glavin and Nowinski threw themselves into it.

Raising money was very disruptive [said Glavin]. I shudder to think of all the time I spent calling around to prospective investors. For every one you get, you have to talk up ten. And those meetings with lawyers. Spending an hour arguing over a word. It's not fun. I spent at least 50 percent of my time raising money. For every deal, you ask lots of girls to dance before you actually get out on a dance floor. It takes you off your focus. We used to take vacations after a round of financing. Get away for awhile.[11]

If Glavin found himself pulled from the operational details most chief executives deal with, Nowinski had less time for the science.

Contractually, Nowinski was supposed to spend 35 percent of his time on the Syva research project.[12] But after road shows, analyst meetings, and chats to potential investors, he had little enough time for research. Nonetheless, the company met its goals in developing the sexually transmitted disease tests.

The real danger of this furious fund raising was overexposure. The Blechs and Genetic Systems seemed always to be selling something new. A blizzard of press releases descended upon the media, analysts, and institutional investors, announcing one deal, one new employee, after another. To some the company seemed to be trying just a bit *too* hard. For outsiders like the Blechs, press releases, like advisory boards, were a surface phenomenon that could be read in many different ways. "Some companies, and Genetic Systems in particular, put out tons of press releases," said one money manager. "If the president sneezed, they put out a press release. It may be dumb but I have a prejudice against companies that do that." There is a thin line that separates information and hype, and the brothers and Nowinski were very close to crossing it.

As George Todaro flew into Seattle, October 1982, Nowinski was working on a new deal. Syntex and Genetic Systems had coexisted successfully for more than a year. Genetic Systems had shipped the first few sexually transmitted disease monoclonals to Palo Alto, and Nowinski was riding Syva hard to package them into kits, do the required testing, and get the data to the FDA, so Genetic Systems could show some product revenues. Nowinski, at his abrasively demanding best, was driving Syva management crazy.

While Nowinski leaned on Syva middle managers, he struck up a congenial relationship with the men at the top, David Rubinfien, who ran Syva, and Albert "Burt" Bowers, the chief executive of Syntex, who had started with the company as a chemist during the glory days in Mexico. Bowers, as Glavin said, "had an inclination for research. Syntex always spent more than nearly anyone else in pharmaceuticals which is a testimony to Bowers. He was a scientist and he seemed to respect Nowinski." In its offering prospectus, Genetic Systems had mentioned using monoclonal antibodies against cancer—first as a diagnostic, then as a therapeutic. This was an oft-discussed possibility, but the project was, in Glavin's words, "squishy. We weren't really sure what we would do."[13]

Now Nowinski began talking up cancer. When he wanted to convince, he could be extremely persuasive. He had several people in mind to run the cancer project, including George Todaro. Todaro certainly knew the cancer field—oncogenes, growth factors, oncostatin—and he had a major reputation. They agreed to meet when Todaro came out to speak to the people at Fred Hutchinson. Over dinner, Nowinski pitched the deal. "There was nothing here. It was really a bit of a dream," said Todaro. "But Nowinski was tremendous at building the dream. He said this project was emerging and how he needed someone in the area of cancer. That you could recruit who you wanted, and do your own things. He was very convincing; he has this amazing capability."

Todaro was ready for a change. What Nowinski proposed was unlike any industry position he knew—certainly not like Merck. This had a university flavor: ties to Fred Hutchinson and the University of Washington, the right to publish, financial backing, freedom. "As it was originally set up, I wouldn't have to concern myself with manufacturing and marketing problems that other biotech companies had to worry about," he said. "Syntex took care of that. I was insulated from it." And Todaro was swept up in the excitement of biotechnology. The traditional barriers between the academic and commercial worlds seemed to have fallen. The merits of the traditional community—the ability to tackle the most difficult problems, the free interchange of results—now seemed possible in biotechnology:

> I was aware that some of the oncostatins, some of the monoclonals, some of the approaches coming out of biotechnology, would result in new pharmaceuticals and diagnostics. I had been to Genentech a few times; obviously they have people as good as at Harvard or MIT. It was a revelation to me: There was more than one way to do basic research and the fact that it was practical . . . well, I liked that.[14]

Todaro expressed some enthusiasm. Nowinski began more serious negotiations with Syntex. It would be located, as Todaro wished, in Seattle, not Palo Alto. The second floor of the Genetic Systems building stood empty. The deal was structured as a joint venture disguised as a limited partnership. Syntex and Genetic Systems served as the two general partners—for all intents and purposes the only two partners—while a limited partner called ONKEM was set up to insure the tax benefits. Syntex agreed to contribute $8 million over a four-year period, with Genetic Systems chipping in $1.5 million. In turn, Syntex would get proportionally more of the early revenues that resulted—unless Genetic Systems desired to even the financing burden.

Todaro also negotiated a four-year contract. He would receive $100,000 a year in salary, as well as moving costs and a mortgage carrying fee for the house in Bethesda. He would get a car. With Todaro as a consultant to Genetic Systems, the company would pay tuition in college for his eldest child; if he remained in that role for three years, it would also pay his younger children's tuition. Nice, but the real difference from NCI came from the stock options: by simply staying on the job, he received options on 7,000 shares in Syntex and 105,000 shares in Genetic Systems, well below the market price. After the first fiscal year in which the venture produced $15 million in revenues—no mention of profit—he could buy 3,000 more Syntex shares and 45,000 Genetic Systems shares. As a member of the scientific advisory board, he got a chance at another 50,000 shares. Most rewarding of all, Todaro was given a 1.5 percent interest in a limited partnership, where he would receive up to 6 percent of profits of the project, if they ever materialized.[15]

CHAPTER 10

The Interpretive Challenge

I N THE FIRST PHASE of biotechnology, 1983 represents a sort of speculative climax. The biomania that resulted from the Genentech offering was marred by a recurrent ambivalence as if there were some warring division between Wall Street's heart and mind. Investors would feverishly buy shares on introduction, then proceed to sell them off; the price of shares would thus quickly climb on the offering, only to sink slowly in the weeks and months ahead. It gave the appearance of great activity, and many companies did come public, but the end result for investors and the firms themselves was often deflating. By 1982 the reaction had set in: investors, battered by the effects of the ongoing industrial recession, had all but abandoned this most speculative of stock groups.

Late in 1982, however, the mood began to change. Investors, buoyed up by a strengthening economy, began to discard their uncertainties. The economic future, in general, appeared brighter. And when it came to biotechnology, the news that made its way into the press and the reports of Wall Street analysts was good. Eli Lilly was trying to get its human insulin, genetically engineered by Genentech, approved by the FDA. Hybritech was actually selling a few of its own antibody-based tests. And then, on January 7, Genetic Systems announced both the formation of its own glamorous cancer research venture and the recruitment of Todaro. The venture's name had a certain resonance:

91

Oncogen—clearly, a company serious about cancer. Analysts began knowledgeably chatting up oncogenes and growth factors and Todaro's own anticancer protein, oncostatin, and on Wall Street Genetic Systems stock surged. By March it had moved some 400 percent, from just over two the previous August, to ten. The Oncogen announcement sent it up two more, to twelve.

Market psychology, like human nature, refuses to be completely elucidated. The crash of 1987 offers just another lesson on that hard truth. In fact, one does not so much "play the market" as play with it—very gingerly. Rational choice may increase the odds, but systems claiming to predict market behavior inevitably crash; that, of course, does not deter folks from trying. Perpetual motion machines attract a lot of attention, too. Hindsight, however, does offer a comforting sense that one can explain market behavior. Beginning in late 1980, whole sectors of the economy, particularly manufacturing, submerged from sight, sapping the strength of the market. Few investors prospered; from November 1980 to March 1982, the market lost $295 billion in value. And as investors headed for cover, they pulled their money from speculative plays like biotechnology and into safer alternatives, blue chip stocks or bonds.

It could not last. Finally, in late summer 1982, the Dow Jones Industrial Index, a basket of large, blue chip stocks, roused itself and lurched forward thirty-nine points. The economy showed life: interest rates and inflation were down, helped by tumbling oil prices, and employment and profits were rising. The long, hot summer of the Reagan years was beginning. Over the next month, the market advanced 150 points. By March 1983 the rally had reached the more speculative technology stocks, particularly the computer companies. Only in late spring did biotechnology as a group begin to advance. By early summer, the initial public offerings were pouring onto Wall Street again.

The optimism of 1983 coaxed forth a variety of exotic corporate life forms. Three days after the formation of Oncogen was announced, a company named Alfacell floated an initial public offering on the over-the-counter markets. Outfits like Genetic Systems tapped the resources of big investors—wealthy individuals, investment funds, and a few institutions. These investors might know little more about biotechnology than they could read in the prospectuses, but they were, in the scheme of things, sophisticated. Many invested for a living. They expected a certain orthodox behavior; they demanded certain results.

This new optimism brought another kind of investor—the rank

amateur—flooding into biotechnology. This tended to happen in every bull market. These were less discriminating investors, by and large; less wealthy, drawn to the excitement of the market, the possibility of a killing, or perhaps the thought of contributing to a cancer cure. Alfacell's initial public offering was underwritten by a small downtown firm called A. T. Brod, like Muller & Company, a distributor and trader of cheaper stocks which sold the units (consisting of two shares of stock and one warrant) to small investors, including members of investment clubs in northern New Jersey and Long Island.

Alfacell was the creation of Emil Szebenyi and one of his former students, Kuslima "Tina" Shogen. They both talked of curing cancer with a substance called NSTT, nonspecific tumor toxic. Dr. Szebenyi— he was a Ph.D., not a physician—was born and educated in Hungary, and the son of a physician. With his wife and three children, he fled the country during the 1956 insurrection and, after some odd jobs, settled in as a biology professor at Fairleigh Dickinson University, in Rutherford, New Jersey, in 1963; nine years later he became chairman of the department of biology.

One day in 1969, Szebenyi reportedly dispatched Shogen, his student, to work on an idea that he had brought from Hungary: a certain biological process that, claimed a public-relations spokesman years later, had been in the scientifically oriented Szebenyi family for several generations. Thus, while working with animal tissue cultures, Shogen reportedly discovered NSTT, which seemed to have extraordinary biological effects. Oddly enough, no papers were written, and no mention of NSTT seeped into the academic world. This would seem to indicate either early commercial ambitions or great uncertainty. To this day, the events under which the discovery were made remain secret, withheld from an outside world, according to Shogen, eager to seize upon NSTT for its own profit or to foul up her already strained relations with the FDA. Shogen presents her own mysteries: often described as an "honors" student, apparently for some student prizes, she never did research outside Szebenyi's orbit, thus escaping the judgment of her peers. Indeed, her scientific reputation exists only because she and Szebenyi say it does.

Szebenyi and Shogen often said that they labored on NSTT throughout the seventies. In 1976, Shogen left Fairleigh Dickinson to form MKS Research, Inc., a "consulting biomedical firm." In August 1981 the pair incorporated Alfacell Corporation in Bloomfield, New Jersey, aided by a former stockbroker named Andrew Oras.[1] At the time, Oras was running a company that sold laboratory animals called Somerset

93

Breeding Labs. He had a checkered background: in 1972 the SEC charged him, and twenty-six others, with selling unregistered securities. Oras accepted the charge without admitting guilt. According to Shogen, she first met Oras when he supplied MKS with lab animals. In 1976, she once said, she hired Oras to do feasibility studies on building a company around NSTT. He, in turn, engineered the formation of Alfacell.

There is to this account a sense of something missing. Oras, according to the prospectus for one of his later ventures, was "self-employed as a management and financial consultant to transportation, recreation, medical insurance and natural resources companies from 1969 to 1980."[2] He became president of Somerset in 1981. A varied background, but what did Oras know about cancer research? And why was a financial consultant selling lab animals in 1976?

For all the nagging questions, Alfacell shared with Genetic Systems certain surface similarities. Alfacell, after all, claimed to have found a cancer cure, and what did anything else matter beside *that?* And it wrapped itself in the cloak of biotechnology. It stamped "biotechnology" upon its annual report as if it were some filing aid to a harassed clerk or, perhaps more precisely, the key to some powerful, esoteric knowledge. The company, however, did not use recombinant DNA techniques to produce NSTT. And although it talked about generating custom monoclonal antibodies for clients, that business never developed. The semantic logic seemed to be that since NSTT was a biological material—extracted from animal tissue—Alfacell could call itself a biotechnology company. Bio equals bio.

Alfacell also had a scientific advisory board. Granted, it had no Nobel Prize winners, and the institutional links were somewhat obscure. But it did boast a veterinarian; a Cooperstown, New York, physician (Dr. Stephen Szebenyi, Emil's son); a veterinary director from Montefiore Hospital; and Dr. Oleh Hornykiewicz, the chairman of the Institute of Biochemical Pharmacology, University of Vienna, and director of the Human Brain Laboratory, the Clark Institute of Psychiatry, University of Toronto. An eclectic group—one lacking anyone who knew much about cancer. Perhaps to supplement those skills, Alfacell's board of directors also featured a New Jersey dentist, Dr. Allen Siegel, and Emil's other son, Dr. Andrew Szebenyi, a Norristown, Pennsylvania, thoracic surgeon. Both received stock in exchange for "consulting" relationships.

Like the monoclonal companies, Alfacell promoted a two-stage strategy. Instead of leveraging diagnostics into therapeutics, it planned to develop a line of biological products which would be sold to fund the testing, purification, and regulatory approval of NSTT. Whereas Genetic Systems highlighted its scientific staff, its skill in antibody development, and its attractiveness to corporate partners, Szebenyi boasted of Alfacell's *animal* facilities. "The Company's environmentally controlled facilities for breeding and maintaining the majority of its laboratory animals," wrote Szebenyi to shareholders, "gives us a headstart in the production of many of our biological products and a flexibility which is rare among biotechnology companies."[3] Mice were important, but they were not that important. Between the vets on the advisory board and Somerset Breeding, animals received extraordinary attention at Alfacell.

Despite that, the "biotechnology division" foundered. In its first annual report, Szebenyi offered a "conservative" projection of $3 million in bioproduct sales in 1984. Sales, alas, amounted to a mere $27,000 that year, rose to over $300,000 in 1985 from one big sale to a cosmetics supplier, and then fell back to about $100,000 in 1986.

Still, talk of a cancer cure, together with the aura of biotechnology, played well to small investors. But if the financing beneath Genetic Systems was complex, that under Alfacell was bewildering. Early on, the company had funded itself by privately placing stock. Then, in early 1983 it began to issue complex combinations of common stock and warrants that clicked off into the future like a long row of falling dominoes. First, it sold units, some 330,000 in early 1983. One Alfacell unit consisted of two shares of stock and one warrant. That warrant, in turn, would allow the holder to buy another share of common stock at $3.00 six months after the offering, or $3.50 after nine months.[4]

That was just the beginning. A few months later, Alfacell offered two more series of warrants. Each series—there were 330,000 warrants in each—could be bought by those who had exercised earlier warrants. The first series, which expired in June 1985, could be bought for $6.50 and would bring in just over $2 million; the second, expiring one year later, in June 1986, was worth ten dollars a share, and would bring in over $3 million.

To many Alfacell investors, NSTT was the ultimate article of faith. The warrants kept them baited financially, while Shogen's oracular comments about NSTT kept them involved emotionally. But while Alfacell skillfully withheld information, bits and pieces of "inside" knowledge percolated through the network of retail investors. Much of this was transmitted by an energetic investor named Martin Blyseth, an engineer from Long Island, who began to write reports on the company for investors. Over the years, Blyseth was alternately impressed and exasperated by Szebenyi, whom he familiarly called "Zeb" or "Dr. Z," and Shogen, "Tina." In 1984, Blyseth wrote:

> In particular, Ms. Tina Shogen, working with Dr. Szebenyi, appears to have devoted her life to the pursuit of NSTT. One can easily conclude this has been done at great sacrifice, personally and financially. In a nutshell, these two scientists believe and are committed. The fourteen years of research, devoted to the delineation and purification of NSTT and their search for the theory of its operation, may well be a story worth telling in its own right. Alfacell is not just another new venture startup with a large staff of Ph.D.'s waiting to start from scratch; we may well be witnessing the successful culmination of a life's work.[5]

Blyseth attempted to place NSTT in a scientific context. He attempted to suggest a mechanism, albeit a sketchy one. His reports had the same combination of reticence and forwardness, naïveté and knowledge, that characterized Shogen's pronouncements:

> NSTT appears unique in that it attacked malignant cells without detrimental effect to healthy tissue. . . . According to Alfacell, NSTT enters the system intravenously, and when it is introduced into the tumor-bearing animal, appears to selectively move to the tumor and carry out the biological actions whereby the tumor cells are literally destroyed. NSTT does not appear to operate in the same fashion as other anticancer drugs. The NSTT appears to be considerably more basic. The tumor cells appear to be "washed out" of the system such that the tumor ceases to exist after a period of time. According to an Alfacell consultant, *there's a quality here that differs from almost everything that he's aware of* with regard to its application to anticancer treatment, and does not appear to need the immune system in its operation. . . . As opposed to chemotherapeutic agents, when viewed microscopically, *it appears to operate at a different level of disengagement;* and appears to be related to a much more basic biological *process*. To date, they have seen no toxic side effects. Of greater interest is the fact that because of the biological nature of the process, as best understood, *there is nothing in this process by which one would suspect that the human event would be different than seen in the animal tests.* [Blyseth's italics]

Blyseth's audience knew little about biology or cancer research. As a result, Alfacell represented a more extreme case of the same interpretive challenge that wracked all of biotechnology. It was easy for a sophisticated viewer to dismiss Alfacell on the circumstantial evidence; the facts were elusive, chimeric, apprehendable only through a maze of explanations that inevitably turned back upon themselves. But for the less sophisticated, the company created an aura that tapped into the potent myth of the entrepreneurial biotechnological company. From 1981 to the end of 1985, Alfacell lost about $5 million. Where did the money go? About a million dollars went toward "research and development salaries." This was a kind of catchall bonus, often paid out in the form of stock. In 1984 alone, $512,940 out of $534,716 of R&D salary expenses were disbursed in the form of stock. Consulting agreements were rife. In 1984, Alfacell set up a stock plan; and in less than a year, the board disbursed shares to two public relations consultants, two other unnamed consultants, three directors, a scientific board member, and three officers, including 30,000 shares worth over $400,000 "to two officers . . . as bonuses for their development of Pannon and the business of the Company"—in other words, Tina Shogen and Emil Szebenyi.[6]

More complex deals took place as well. Soon after, Alfacell expanded to a second facility owned by Shogen. Then there was Pragma Biotech, a company controlled by Oras, who had served as Alfacell's vice-president. When Pragma was founded, he decided to leave Alfacell. He did not go far; Pragma's offices were at the same address. As payment for allowing Oras to leave, Alfacell received stock in Pragma, which was trying to develop something called diagnostic organ specific cancer indicators—DOSCI for short—to replace tests for a common cancer marker called CEA tests. Its prospectus talked about using monoclonals to target tumor antigens. Not unusual. Unfortunately, by late 1986 Pragma admitted to having made "no progress" on the tests since October 1984. In fact, by September 1986 it "recommenced" development by paying a "PhD" $11,000 plus expenses to generate a *polyclonal* antibody. Pragma also talked about setting up a cancer testing clinic on an "as yet to be determined Caribbean island." Soon after, Pragma won the contract for phase-one testing of NSTT from Alfacell, a deal that was steadily sweetened over the next year. Most of the money was funneled to Pragma Dominicana, a Dominican Republic outfit that would actually do the tests. Oras and Julius Isman, an Alfacell and Pragma board member, served on Pragma Dominicana's board, but both denied any equity or compensation in it.[7]

While all this was taking place, Alfacell was moving to test NSTT. In late 1984 it renamed the compound "Pannon," and took it out as a trade name. "A question often asked is, 'What is the origin of the name PANNON?'" wrote Blyseth. "We finally got to ask Ms. Shogen and were surprised to find that its origin was a combination of (1) 'pan' as in all-encompassing and (2) 'Pannonia,' a Roman province encompassing part of western Hungary. So what we have is a name whose source has a flavor of both the ethnic and the optimistic. Most of those we asked guessed 'panacea,' which one day may hopefully be appropriate." As usual, Blyseth finished his report with an adage for the doubtful. *"Lest we forget:* This is a story of which fairy tales are made . . . and dreamers rewarded. . . . So, *don't close your eyes to the risk,* only to the dream."

By mid-1984, Alfacell, like Genetic Systems, had thousands of warrants floating around. Much that subsequently took place has to be viewed in relation to those warrants.[8] After fifteen years of work on Pannon, Szebenyi and Shogen only began to discuss human trials after the stock offering, in 1984, and the looming presence of those warrants. Over the next two years, the company, which, despite its public-relations fees, always claimed to avoid the press, would begin a rising crescendo of announcements in January, driving the price up through winter and spring. These releases were accompanied by a haze of rumors, gossip, and speculation. Finally, the price would pass the warrant threshold, and holders would shell out for common stock. Money would pour in; announcements would slow; predictions would be hedged; the press, nefarious traders, and even shareholders themselves would be excoriated for their irresponsibility. And the price would slide.

It began with the 1984 warrants. In April came Blyseth's first major report. He argued that if the warrants expiring in May were not exercised, "Alfacell would have to seek alternate financial support to begin human clinical testing." But there was a bright side:

> If the remaining (85%) warrants are exercised they should bring in just under $1 M. This would be sufficient to begin initial testing, and continued success will give the company the leverage to raise the additional $3 to $5 M that may be required towards the end of 1984 and into 1985. Early success will bring not only easy access to capital but the possibility of the FDA as a partner on the FDA fast track approval route.

The stock did rise and the warrants were exercised on June 25, giving the company over $900,000.

Still, the company did not enter the clinic, did not raise more money, and did not have any visible effect on the FDA. The stock fell back to four. In August, Blyseth addressed the impatience of investors. "Believe me," he said, "no one appears to be more eager to proceed than Alfacell principals." Besides, testing preparations had begun. "Under the co-ordination of Roger Fidler [Alfacell's patent lawyer and vice-president

The Interpretive Challenge

for corporate planning], patients are being screened and the first group of five or six have been selected. Fifteen to sixteen such groups will be processed. Among other criteria, these patients have been selected based upon the fact that they have not yet responded to other treatments, but are in relatively good health." Testing would be done in the Dominican Republic clinic of Dr. Angel Chan Aquino. That country had applied to the FDA to export Pannon to the clinic—and the testing should have met all FDA phase 1 and phase 2 requirements.

Although testing did begin in November, the sixty or so patients soon became "four [patients] building to sixteen"—no groups; no mention of Pragma Dominicana; No names of any clinicians beyond Szebenyi and Aquino. Nonetheless, in December, Shogen announced that the pair were "very optimistic" after preliminary results. Rather than perform the phase 1 toxicity tests on healthy persons, Shogen said that an institutional review board "supervising the trials" had "decided that humans afflicted with cancer should be used . . . in the hope that the cancer sufferers might be helped, with even a minimal dose of Pannon."[9] At the annual meeting a few weeks later, Fidler added that the company would not seek additional financing until conclusive results were in, by January or February. Actually, the company did seek financing in the form of a bank loan. In early February the company announced that "the review boards" had decided to merge phase 1 and phase 2 protocols—testing for safety and effectiveness—essentially what they had already done in December.

Blyseth took that as an optimistic note. He was now openly comparing Pannon to tumor necrosis factor, a lymphokine like interferon which, alas, proved to have serious side effects, and Alfacell to Genentech: "This may be the most exciting FOOTRACE of the century," he wrote. Expect test results, he reassured investors, in another sixty to ninety days. And don't forget those warrants, due in June. By the end of March, Alfacell stock had risen to twelve, in furious trading. Rumors flew. The *Professional Tape Reader*, a Florida newsletter, included Alfacell among twenty-five other "interesting" stocks. That assessment was soon reported as an outright recommendation by the *National OTC Stock Journal*. Around the same time, Shogen announced that "prestigious" investigators would soon visit the clinic. The rumor mill soon had the FDA sending investigators to the Dominican Republic. In early April the stock hit fourteen. The *Portfolio Letter*, another newsletter, quoted Shogen as being optimistic about results, while refusing again to name the investigators so to protect them from being "bombarded by anxious shareholders." The newsletter also reported that "street sources estimated that a report [on the tests] should be out in ten days." Much was made of her desire to protect Alfacell's good relationship with the FDA. Finally, in late April, with the stock into the twenties, came a negative note. *Barron's* discovered that the FDA had turned down the request to import Pannon to the Dominican Republic for lack of information. Instead, Alfacell exported the raw materials to Canada and shipped Pannon from there, slipping around the regulatory authorities.[10]

No matter. A week or so later, the company announced "an interim report on the well-controlled trials ongoing in the Dominican Republic in human cancer patients having the following carcinomas: colon, stomach, rectum, breast, larynx or penile indications." The bombshell, ticking over the Dow Jones wire with quarterly earnings reports and stock prices, came in the second paragraph. "The clinical testing has proved that Pannon destroys malignant tissue without affecting healthy tissue."[11] This certainly sounded like a cancer cure. The tests showed "that Pannon is more effective in humans than in animals permitting lower daily doses . . . to dissolve the tumors." Alas, a few days later, the FDA moved to stop shipment of Pannon. A difference of opinion, said Fidler, adding the company had no need to ship more anyway. The company also announced that it would present its findings to the FDA in August. By June 25, with the stock back down to fourteen, more than 95 percent of the $6.50 warrants had been exercised, and Alfacell had almost $2 million in the bank.

In September 1985, Alfacell had still not submitted its results to the FDA. But it did hold a meeting for investors in a smoky, hot room at the City Mid-Day Club on Wall Street. The room was jammed with investors, in from Long Island and New Jersey, retail brokers, and a few journalists. The investors were restless. Shogen, a thick woman with a school-teacherish air, tried to calm the waters. She talked on about the grand success of the trials, about her optimism and faith. One by one, her officers stood up to testify: Fidler; Joseph Barrows, a consultant for regulatory affairs; the new medical director, Stanislaw Mikulski, who struggled in thickly accented English to explain hazy slides of cancer patients and unfathomable charts demonstrating that Pannon inhibited DNA synthesis in culture; and Steven Carson, a toxicologist from St. John's University. On a table rose thick, blue binders: the test data that were about to go to the FDA, said Barrows. The crowd, however, wanted more tangible results; they had put their hard-earned dollars into Alfacell. They were not professionals; indeed, there was no major institutional presence at all. They were just ordinary, middle-class folks who thought they had stumbled upon a grand opportunity to eliminate cancer—a scourge many had known first-hand—and make a profit. Now they were confused; they were not sure who to blame; they were beginning to feel victimized.

By then the company had discovered a new bogeyman: the short-sellers, a class of professional investor that makes money when a stock falls—or is driven down—instead of rises. Clearly, short-sellers were active in Alfacell. But were they justified? Was Alfacell a company with a bright future? Was Pannon a true panacea, or a product of smoke and mirrors? Who could tell, although the shorts bet the latter. And as time went on, Alfacell proved unable to counter the doubt and suspicion that the shorts feed on. In November the company finally filed an Investigative New Drug (IND) application with the FDA which, it said, would allow it to conduct human trials at "a number of prestigious institutions in the United States and abroad." Alas, by the time the annual meeting came in March, Alfacell claimed that the FDA had transferred Pannon from the Center for Biologics to the Division of Oncology. Companies usually file an IND and wait a month; if they have not heard from the FDA, they begin trials. Here it was five months later and nothing. Indeed, the company was still waiting for its IND more than a year later, in December 1986.

This time Alfacell was unable to lift the stock off the ground and redeem those warrants. By the end of 1985 it had fallen to eight. Around the time of its meeting, it struggled to almost fourteen, before eroding again. The company again blamed the shorts. In September, Blyseth produced his last written report, still upbeat, but passing along the unfortunate news that Alfacell had gone back to *animal* testing and was further purifying Pannon. By then the financial situation had grown even more serious. The company had taken out more bank loans; in December it reported $800,000 in short-term debt. The company extended the warrants. Szebenyi then retired, shuffling off the stage. In September, Shogen sent a letter to shareholders, "the first of an anticipated series of communications." Don't worry, Shogen wrote, despite the *appearance* of difficulty, everything is moving along smartly. The company is still testing Pannon, upgrading its facilities, meeting with NCI, and expanding its bioproduct line. An amendment to the IND would be submitted at the end of October. The problems that *seem* to exist can be blamed on the shorts. "The artificial atmosphere being maintained by outside market influences has *no relevance* [Shogen's italics] to the qualities, properties and potential of PANNON."[12]

Perhaps. But short of taking her word, how was anyone to tell? Alfacell eluded interpretation. The unusual operating style, the aggrandizement of insiders, the lack of real qualifications—none of that would have mattered if Pannon had ever received the imprimatur of anyone not associated with the company. If only they could find someone to validate the tests. But Shogen argued that to do that would be to give away the secret of NSTT. Thus, the history of Alfacell became one of denial and delay. Even Shogen's last big promise, that an IND amendment would be filed in October, was altered first to December 1986, then to some time in early 1987.[13] A purified form of Pannon, now renamed P–30, took until March 1989 to enter trials—for which top management gave themselves bonuses. And if those trials fizzle? The magic elixir, Pannon—the cancer cure, the dream—would still exist as an idea, as a possibility. The search for scapegoats could then begin in earnest: the shorts, the big drug companies, the FDA, the NCI, the cancer establishment, or some investors themselves.

In time, investing in Alfacell was more a matter of faith than rational belief. Alfacell required true believers. Like any closed system, failure to believe fully made one a *persona non grata*. If you are not for us, you must be against us. Your doubt and skepticism are a contagion and might be the very problem. Again, although Alfacell and other marginal operations made these kinds of assertions more blatantly, mainstream biotechnology was not all that far off in its demand for optimism. Biotechnology, after all, was a revolution; and this was a Manichean struggle for hearts and minds. The complex, or partial, view was dismissed.[14]

The interpretative challenge of Alfacell also plagued biotechnology as a whole. The burden fell upon the research analysts of Wall Street. As a group, analysts are hired to ponder, to analyze, to offer up considered opinions. They write reports. They crunch numbers. They construct models as intricate as Tinkertoys. They are expected to be right, but in the universe of Wall Street, they are expected to be something else as well: smart, shrewd, *timely*. Analysts are most intimate with the phenomenological gap between corporate reality, on one hand, and the gyrations of stock prices on the other; many analysts have found themselves irretrievably lost, like wayward cowboys, in the wasteland between the two.

In the world of Wall Street status and power, analysts fit somewhere between the brokers below and the deal-making investment bankers above. They suffer from divided loyalties. They must wrestle with a market they can influence but never control. They must glean data

The Interpretive Challenge

from companies that may dissemble, distort, or lie. They must contend with bosses who know that a glowing report may attract investment banking business—and since brokerage commissions were deregulated in 1975, investment banking has increasingly brought in the profits—and with institutional clients that demand objectivity. They must remember that everyone profits from an up market; negativism may sell stocks over the short term, but it can threaten one's job over the long term. Consistent sell recommendations aren't good for business—a tacit admission of misjudgment.

Biotechnology only added to these burdens. One day in 1986 I sat in the office of a biotechnology analyst on Wall Street. It was late in the day; the markets had recently closed and outside dusk fell. The analyst, an exceedingly astute man, sat in the green glow of his Quotron terminal. Above him ran a shelf stuffed with economics, accounting, and biology texts, government reports and marketing studies. How, I asked, can anyone evaluate companies lacking products or profits or, often enough, even sales? "It's not easy. A lot of what we do is to provide a rational basis for instinct," he said. "We provide a sort of window dressing for investment. And we try to educate people at the same time."

A few minutes later he rose to leave to talk to the brokers over an in-house intercom system. When he returned ten minutes later, he was trailed by two brokers, eager though wary. They wanted information so they could talk more effectively on biotechnology to their clients. "Something," one said, "that explains what biotech really is."

"A book?" he said hopefully, reaching for a thick volume from the shelf. He laid it down: *The Biology of the Cell.* They stepped back.

"Have you got anything simpler?"

He thought a moment. Like every other biotechnology analyst, he had written a primer on the technology and the industry. Biotech analysts put out their shingle by explaining biotechnology to unwashed investors. It defined things like recombinant DNA, antibodies, and antigens. It had simple drawings of cells, double helixes, and antibodies. He reached into a cardboard box and pulled two out. "I don't know," said one of the brokers, squinting at it. "Looks complicated. Don't you have anything simpler?"

"It doesn't get much simpler," he said.

They left, clutching their materials like a bomb that might go off. "If you think it's bad now," he whispered, "it was worse a few years back. Now, at least, they ask." He fell into his chair, his eyes seeking the solace of the Quotron. "It makes this job," he said, sighing, "more of a challenge."

In fact, biotech analysts, particularly in the early days, always operated a bit outside the analytical mainstream. The job was new, many of its practitioners were drug analysts just learning molecular biology, or refugees from academia, just getting a feel for Wall Street. A few were not very good in either sphere. Biotechnology analysts differed from their colleagues in the same way that biotechnology differed from more traditional industries. The analysts had plenty of tools but none that particularly applied to this problem. It involved an abstruse subject that would not generate earnings, the basis of fundamental analysis, for years. Thus, institutional clients, by necessity, had to lean harder on biotech analysts than others; they required not only advice, but education. That relationship cut both ways. On one hand, an analyst could be wrong for a long, long time before anyone caught on. On the other, the speculative gap made clients jumpy, liable to flee at the first ominous sign. Piled atop the usual analytical problems, biotech analysts had to sell the *idea* of biotech, inside the firm and out, if only to justify their own jobs.

All this helps explain why Genetic Systems could trade at two one month and at eight four months later, or why an Alfacell could rise and fall like a yo-yo. Investors were nervous, liable to sell out over the slightest unhappiness. This nervousness was hardly the only factor involved in volatility; larger market forces, small floats, and the economy all played a part. But all these pressures converged upon analysts.

There is a theory, often heard on Wall Street, that the operation of markets is a sort of information system. Input in the form of data comes in one side, investors make their choices, and output in the form of pricing goes out the other—a simple mechanical metaphor in which pricing is based on the ability of investors to react *rationally* to information. What makes Wall Street a game, with winners and losers, is the degree to which data are ever completely absorbed or fully comprehended; that is, in Street terms, these markets are never fully efficient. Plugging biotech into this makes for a highly inefficient machine. Molecular biology was a body of information forming incomplete, perhaps deceptive, patterns; like the market itself, one could argue about many conclusions. If eminent scientists could not agree, how could analysts negotiate the currents? And stepping back: How could brokers, sweating over simple drawings of DNA, or portfolio managers, eager to invest billions, place their clients' money rationally? How could doctors, dentists, and members of investment clubs do so?

CHAPTER 11

Inside the Mouse Factory

BY THE SPRING OF 1983, Genetic Systems was riding high. The market was rising; a whole raft of products, including the sexually-transmitted-disease tests, seemed imminent; Oncogen gave it an entrance into cancer therapeutics. Although the company was still losing money, that was to be expected. The company had been operating for two years; it was, compared to many others, a biotechnology veteran. Imagine the progress it would make in the next two.

Opportunity beckoned for a second public offering. D. H. Blair was out; the big guys, Allen & Company and Merrill Lynch, now appeared to manage the offering. This time, there was no need for warrants to sell the stock. On April 7, the two brokerage firms pushed 2.2 million shares onto the market at about ten dollars a share. All went smoothly, carried along on a bubbly, optimistic mood. Champagne was uncorked. The company now had 18 million shares outstanding, with a market capitalization, at a price of ten, of $180 million.

The logic of venture investing was spelled out in the prospectus. Founders, officers, and early investors owned about 20 percent of the company but had paid in only $45,000, or about one-tenth of a percent of total shareholder capital. Their stock, at ten, was worth about $36 million, 800 times their original investment. Those who bought at the initial offering or converted their warrants into stock had contributed about $15 million, or 41 percent of capital. They owned about 68 percent of the shares, worth $130 million—almost nine times their original stake. And the new investors? They were paying in almost $22 million, or 60 percent of capital, for 12 percent of the stock. They had no profits yet, although the future beckoned.[1]

Numbers like these could turn the head of the most hardened Wall Streeter. This is how the game was supposed to be played. The lesson was: Buy now or pay a higher price later. Indeed, before the summer was over the stock would go as high as seventeen.

Genetic Systems' apotheosis may well have been in 1983. With a big public offering, a limited partnership, corporate deals all over the map, $36 million breeding interest in the bank, and $15 million more in Class B warrants expiring in June of 1984, the company had acquired financial muscle. And, with Nowinski, Todaro, and the likes of Dr. King Holmes, the noted expert on infectious diseases, they had scientific muscle as well. And, more tangibly, products were finally appearing.

Syva had packaged most of the sexually transmitted disease antibodies by now, run clinical tests, and gotten FDA approval, and it was just beginning to market them. The sight of products brought forth the analysts. Nina Siegler, an analyst at PaineWebber, in an impressively detailed report, saw Genetic Systems "turning the corner in 1984," with earnings of five cents a share and revenues of $12.5 million.[2] A few months later, Nelson Schneider wrote, "Genetic Systems appears to be the third company in the biotechnology field to be in line for significant product sales and earnings in 1984."[3] He estimated revenues of $20 million and earnings of fifteen to twenty cents a share. Journalists wrote stories such as, "Antibodies Hold Big Profit Potential," or "Unlocking Cells: Monoclonal Antibodies Hold Vast Medical Promise." One did not hear much about interferon anymore. Alpha interferon had proven itself safe and was into the final stages of clinical testing, but the big question—does it work?—remained unanswered. Mathilde Krim's interferon lab at Memorial Sloan-Kettering still operated, but the spotlight had swung away. Scientists were leaving. And there was talk of another medical crusade: a mysterious immune disease killing homosexuals, Haitians, and hemophiliacs.

Besides, the hot game in 1983 was monoclonal antibodies. And the key to monoclonals, for the moment, was diagnostics. A great little business, diagnostics; a great little money maker, too.

Inside the Mouse Factory

*   *   *

For all its importance, diagnostics never got its due. Diagnostics lacked class. It sounded like something you did to your car. "Let's check out the carb with the old Diagnose-a-meter." Or it evoked memories of marriage licenses and rabbit tests. Companies selling diagnostic tests couldn't brag about *curing* anything, although they might make a cure possible. Even for the companies themselves, diagnostics fell into the strategic limbo of means rather than ends. The strategy of using diagnostics to break into monoclonal-based therapeutics was like taking a summer job to pay for Harvard in the fall—a prelude to greater glories. The rich kids on the block, Genentech, Cetus, and Biogen, did not in those days take either monoclonals or the diagnostic business particularly seriously, although they did, on occasion, dabble in it. And Wall Street never offered diagnostic makers the same sorts of stock premiums as on those of the major recombinant firms. You did not succeed in diagnostics with one home-run product; it required lots of singles. Babe Ruth always got more votes than Ty Cobb.

It was, in fact, a mouse game, a mouse-intensive industry. Nowinski would jocularly refer to himself as "the mouse doctor," and Genetic Systems, like other antibody companies, was a mouse factory. "For fifteen years my work has been in mouse biology," Nowinski was once quoted as saying. "I used to milk mice. And do mouse surgery. I'm known as the 'mouse doctor.' "[4]

Mice were the engines of the business. Mice provided homes for hybridomas, the half-cancer, half-antibody-producing cells that Milstein and Köhler had invented. Genetic Systems, for instance, needed an antibody that recognized the chlamydia bacteria—and only that bacteria. To get one, it had to infect a mouse with the bacteria. As the mouse battled the illness, its spleen churned out B-cells which, in turn, pumped out antibodies that recognized the antigens of chlamydia. Researchers then removed the spleen, minced it up into tiny bits, mixed it with cancer cells, and added a chemical that induces some of the spleen B-cells to fuse with the cancer cells. The rapidly growing fused cells are isolated. Then, careful as a gardener planting tulip bulbs, the researcher places individual hybridoma cells in the abdomen of the mice, letting them grow into tumors. Eventually the tumors leak monoclonal antibodies into a fluid that fills each mouse's chest cavity, which is then drawn off, and testing begins.

If a monoclonal lab has fifty people, twelve may clone hybridomas and the rest test steadily, a dull but absolutely essential task. The fluid

107

removed from the mice is placed in microtiter wells—rows of shallow indentations in plastic trays—with a pipette. In the bottom of each well, different antigens, say for chlamydia, are bound to a plastic matrix. After incubating for an hour or two, the antibody is washed out, and a different fluid is pipetted in. This contains a second antibody, tagged with a fluorescent or radioactive marker, that binds to *any* mouse antibody. In those wells containing antibody sensitive to a chlamydia antigen, a so-called molecular sandwich forms: on the bottom, the chlamydia antigen; in the middle, the first binding antibody; on top, the mouse antibody with the tag. The wells are washed again, leaving behind, like seaweed clinging to seaside rock, the linked molecules topped by the fluorescent markers. A technician can then flash a fluorescent lamp from well to well, searching for the telltale glow of antibody sensitivity.

This is just the beginning. Many questions have to be answered: How great an affinity does the antibody possess for the antigen; how strong, in other words, is the binding? Is it a class of antibodies that are all but impossible to deal with, such as the immunoglobulin M's, a good percentage of everything cloned, or is it the more usable immunoglobin G? Is it an antibody that can be produced in quantity? All this takes thousands of mice, thousands of trays, and months of technician time— one reason so many antibodies came from postdoc-heavy university labs—with one goal: to extract a usable, workable antibody that binds sturdily to a chlamydia antigen, and only to that antigen. In theory, with enough money, enough time, enough mice, researchers might find that one-in-a-billion perfect monoclonal. And indeed, as time went on, as more and more groups cranked out antibodies, it became apparent that *someday* a lack of antibodies would hardly be a problem. But when? Time, alas, was an academic luxury not available to companies with stockholders to satisfy.

Ironically, monoclonals were, in terms of process, not all that different from the empiricism of drug industry practices. The search for the perfect monoclonal only really differed from mass screening because it could refer to a fairly well accepted model of antigen-antibody affinity. But while that principle could be directly applied in the case of certain diagnostics—particularly in the case of, say, pregnancy, when there is one fairly widespread antigen—the gap was much wider when it came to diseases such as cancer. And while the antibody firms tended to view their own virtues as ones of speed and genius, in reality they also had to slog along a towpath carved by mass screening. Instead of chemical compounds, they screened antibodies. Instead of thousands of petri dishes, they juggled mice. Monoclonal companies bought hurrying, scurrying, red-eyed mice by the truckload. So it was no surprise that Bob Nowinski took to calling himself the mouse doctor.

There were many subtle ins and outs to antibodies, however; the theory was simple, the practice was not. Sometimes, using monoclonals for *in vitro* diagnostics could be, from a commercial standpoint, like killing cockroaches with an Uzi: not only was it overkill, but it might not work as well as a can of Raid. For all the talk of monoclonals, old-fashioned, impure lots of polyclonals could prove just as effective, sometimes even better. A few of the big companies knew that. So did a few of those smaller companies that lacked the high-powered science of a Genetic Systems or Hybritech. One of them was run by one Dr. Sigiloff Ziering, whose Los Angeles–based company, Diagnostic Products, was a small, obscure diagnostics supply house that had been making money since the early 1970s. In short, it was one of those companies the biotechnology juggernaut figured to flatten. "The magic bullet approach held people entranced," said Ziering years later when Diagnostic Products was large enough to begin buying ailing biotech companies.

As a result, they used their monoclonal technology on less sophisticated products—ovulation, pregnancy, sexually transmitted diseases. In most cases, these monoclonals are just another class of antibody. For instance, monoclonals sometimes pose a basic problem: they have a lower affinity than polyclonals. A polyclonal may have more binding sites available and so the test works faster. Three years ago, we developed a kit for human growth hormone. The monoclonal would have taken four hours to incubate, the polyclonal only half an hour. Every other characteristic was the same.[5]

Diagnostic Products survived by adaptation. Ziering found niches and occupied them. "I'm not knocking monoclonals," he would say.

For specific kits—sexually transmitted diseases being one—you've got to use a monoclonal. But monoclonals quickly become commodities [in the mid-1980s]. You can take any undergraduate now and teach him how to produce a monoclonal. And the cost difference between polyclonals and monoclonals is not that significant. And I'm not knocking biotechnology. The promise of genetic engineering is very intriguing, but there was a lot of hype, particularly on Wall Street. It was a situation that fed on itself. Wall Street loves the excitement of thinking it knows what's going to happen over the next ten years. It got a little away from reality. Companies got caught up talking to analysts; analysts got caught up talking to them. It fed on itself. If you look at the financial reports, few classified us as a biotech company although we made monoclonals. That's just as well. If they did we'd stand out like a sore thumb: we've been earning profits for fourteen years. We're building a diagnostics company. You won't hear us talking about becoming some big, biopharmaceutical company.

Nowinski felt differently. He realized that diagnostic antibodies did not have to be as perfect as those used as drugs, which meant that one did not have to hunt forever seeking the perfect one. Besides, diagnostics would only be a proving ground—and a source of cash—for the assault on therapeutics. By that time, one could come up with better monoclonals. Besides, everyone knew that the diagnostic industry was relatively new and fast growing, with huge profit margins, fragmented into small companies—400 at one count—many of which were undoubtedly *entrepreneurial*, but like Diagnostic Products, not really *biotechnologically* entrepreneurial. Companies like Genetic Systems would cut a swath like the big companies, with the scalpel blade of the small.

And they seemed to be proving it. By the end of 1982, Hybritech had eight monoclonals approved for sale as diagnostics. Two other companies, Centocor and Monoclonal Antibodies, each got a pair of antibodies approved. Genetic Systems' chlamydia monoclonal proved trickier. As chlamydia travels through the body, it alters its antigens. To develop a diagnostic that picked up the bacteria, Genetic Systems had to assemble an antibody "cocktail," a mixture of antibodies that responded to the variety of antigenic disguises—ironically, a polyclonal mix of monoclonals. This was difficult, but not impossible. By December 1983, two years or so after the company went public, Syva received FDA approval for the first Microtrak chlamydia test, and Genetic Systems, in return, began receiving royalties—a nifty technical feat. The test was still based on culturing chlamydia microbes, but it cut the time down from six to two days. A direct test—a dab of specimen on a slide, a bit of the antibody with the fluorescent markers, see what color it turns under a fluorescent microscope—was also on its way. So were monoclonals for herpes simplex I and II.

By 1982 market forecasters were tossing out new numbers: about 5 percent or so, they said, of the estimated billion-dollar American diagnostic market had become antibody-based tests called immunoassays.[6] That segment was also growing fast, 20 to 25 percent a year. Immunoassays obviously represented the future. They worked swiftly, were easy to use, and could be very precise. And most immunoassays could be fitted into simple *in vitro* test kits, with all the advantages, particularly regulatory, that that implied. The long wait, which really had not been all that long, seemed over. "These areas of biotechnology [monoclonals and diagnostic probes] should no longer be considered blue sky," wrote one analyst in early 1984. "The revolution has arrived!"[7]

CHAPTER 12

The Conservative Reaction

REVOLUTIONS, like bullets, magic or otherwise, have a way of going awry. The theory behind antibody-based diagnostics was wonderful; the numbers, of course, looked fantastic; the future appeared bright. All that, alas, was before anyone in biotechnology tried to actually sell anything. The realities of the marketplace, particularly the competition from larger, nonbiotechnological companies, proved far more difficult than any scientist, locked in a laboratory, could imagine.

In all the excitement about biotechnology, the large drug and hospital-supply companies were rarely mentioned. While several of them had moved to license biotechnology products, notably Lilly with human insulin and Schering-Plough and Hoffmann-LaRoche with alpha interferon, the others were widely viewed as having missed the boat. For the most part, the drug companies continued to wait and watch, unwilling to believe the millenial pronouncements of biotechnology. This, among the biotechnology fraternity, was taken as evidence of the sort of blindness that would, in these fast-moving times, quickly transform the drug companies into corporate dinosaurs.

The large, established companies were, undoubtedly, conservative and essentially bureaucratic. Part of that, of course, was the natural result of size and success; current product lines were so profitable, and life cycles so long, that there was little incentive to launch expensive gambles on untried research. In fact, in the 1970s many drug companies had opted for the safer, if potentially less lucrative, diversifications into cosmetics and consumer goods. That flight toward safety, to be sure, was a trait shared by many other American corporations. But there were other factors that reinforced this conservative tendency. The larger environment that the drug companies operated in demanded a certain conservatism. Medicine and pharmacology were, as we have already seen, very inexact sciences. As a result, the regulatory environment, dominated by the FDA, was extremely cautious; and as the complexity of the science, and the demands made upon it, increased, it was becoming ever more so. Thus, the bureaucracy of the FDA and the bureaucracy of the drug companies mirrored each other. Drug executives dealt with the FDA gingerly, engaged in an ongoing dialogue that consisted of unwritten protocols and seemingly esoteric procedures. The requirements of the FDA, in turn, demanded a commercial organization that had the resources and time to accumulate vast amounts of clinical data and to communicate it effectively—in other words, a highly organized corporation with enormous resources. The redundancies of the bureaucracy were not just an escape from a threatening world; and the conservatism of the industry was not simply the result of years of huge profits and layers of self-satisfied managers.

All this suggested a series of questions that executives within the industry had to grapple with in one form or another. How much certainty, or uncertainty, underlay biotechnology? How important were the strengths of the large companies—clinical and regulatory experience, market presence, monetary resources—in relation to the assets of the small—speed, entrepreneurial fervor, and a technological lead? How hidebound were the large drug and hospital-supply companies? Were the large companies capable of reforming themselves quickly enough to respond to the challenge of this "revolution"? Was there a revolution at all?

In the early 1980s some of these questions would begin to be answered. The clearest, most immediate and most devastating response would come from a $3 billion corporation based in the wooded countryside north of Chicago called Abbott Laboratories. Nearby were the corporate headquarters of other drug and hospital-based firms: G. D. Searle (now a part of Monsanto) and the stunning glass home of Baxter Travenol, with the most complete product line in the hospital supply business. When you arrive at Abbott Park, the word *ruthless* does not immediately come to mind. Despite the security, the complex of woods, parking lots, and chocolate-brown buildings resembles some agricultural college—Abbott A&M—where Ping-Pong passes for a social life. Abbott employees wear suits, ties, and cardigan sweaters. Most reside in the Chicago suburbs, raising families, swinging down to Chicago for dinner, some drinks, and a Cubs game. Mostly, they drive sedans and station wagons; maybe, if they get a little reckless, they buy a Chevy Blazer.

Abbott is a conservative place; the code of conduct is not overly constricting, but few challenge it and remain. There exists throughout the company a sort of professional esprit de corps that, when verbalized, descends into management babble. You do what it takes. Act like a professional. Know yourself, the market. Do the homework. Play to win. Abbott lives physically and emotionally near that Rockwellian utopia, the Heartland, Chicago; it shares neither the edgy cynicism of New York and Cambridge, nor the blithe, sunny confidence of the West. At Abbott those clichés took on flesh.

And yet, to hear its competitors talk, Abbott belonged in corporate reform school. *Ruthless* and *bastards* were terms often combined, as in: "Those ruthless bastards from Abbott." Abbott, admittedly, achieved this reputation through brains and hard work. One telltale sign was that corporate recruiters loved to get their hands on Abbott alumni; the biotechnology industry, for instance, was loaded with them. George Rathmann, the long-time scientific director and president of Amgen, had been an Abbott researcher; Kirk Raab, who had risen to number two spot at Abbott, became president and chief operating officer of Genentech in 1985; and James Vincent, one of the major players in the Abbott diagnostics story, took over Biogen in 1985. Even middle managers were sought. In 1986, Hubert Schoemaker, the president of Centocor, boasted of two marketing managers he had spirited from Abbott. "Tough," he said proudly. "They're very tough at the point of sale."[1]

In this Abbott resembled IBM. Like IBM, Abbott was viewed as a sort of enforcer, a corporation unfairly bullying smaller competitors. In 1984, Hybritech sued Abbott for antitrust violations, charging that it had cooked up a strategy designed to drive it out of business. Abbott, Hybritech claimed, cut prices, bad-mouthed its products, and cannibalized its own analyzers to prevent Hybritech tests from being run on them. Abbott was even accused of having a "war room" to coordinate its campaign. Some of this has the gossipy ring of truth. Abbott apparently did launch an operation against Hybritech's Tandem diagnostics line, which had been designed to run on Abbott's Quantum analyzer by hooking a small electronic device onto it. Abbott ordered its salespeople to rip out the ports, forcing Hybritech to develop, at considerable expense, its own analyzer.

But was this unfair? Was this *illegal?* Hybritech obviously thought so, although the courts have not yet decided—and antitrust suits tend to drag on like Dickensian chancery cases. Beyond the merits of the suit, Abbott clearly did, by its very success, alter the nature of the diagnostics business. "It used to be a very nice, polite business," said Wayne Fritzsche, a consultant and formerly an Ortho Diagnostic marketer; he had also been the head of marketing at Cytogen, a job that abruptly ended when Bob Nowinski jumped to Genetic Systems. "Everybody used to make money. Then Abbott arrived, and suddenly everybody was going around trying to kill each other."[2]

Abbott had been solidly, if unspectacularly, successful for years. It sold everything from pharmaceuticals to nutritionals like Similac to Murine eye drops, and it had been a very long time since it had had the kind of financial problems that faced a company like Genetic Systems. Still, Abbott had two weaknesses that were key to its later aggressiveness. First, it was not a leader in the drug business. Compared to a Merck or a Lilly, Abbott had an older, second-class product line. Second, Abbott had recently undergone difficulties. In 1970 the FDA ordered saccharin off the shelves—and Abbott was the major producer. One year later, in 1971, a contamination scare forced a recall of its intravenous solutions sold to hospitals; although it returned to the market, it had given up its dominant position to its archfoe, Baxter Travenol. Its relatively small diagnostics business, once a leader in tests using radioactive markers, was still profitable, but slipping; there was talk that Abbott might get out of the business, particularly with radioactive tests rapidly being replaced by those using enzymes or fluorescence. The company was struggling to regain its momentum and its reputation.

No one could imagine that diagnostics would lead the rejuvenation. But in 1973, Robert Schoellhorn came to Abbott from Lederle Labs, American Cyanamid's life-science division. He became head of the hospital division, which included diagnostics, and he was directly in line to take over from Chairman Edward Ledder. Reporting to Schoellhorn was another new face, a rarity at Abbott, which traditionally recruited from within: a former semiconductor salesman named James Vincent. Vincent, a barrel-chested man with a thick, square head, had been the chief of Far East operations for Texas Instruments, the big semiconductor and computer maker. But in 1971, bored with chips and stifled at Texas Instruments, he quit. He was thirty-three years old. He sat back to review his options. "Vincent is a very analytical guy," said a former employee. "He must have thought about healthcare for a year before he decided where to go. At first, he couldn't even get in the door at Abbott. Finally somebody got him in, and he showed up with this big, thick business plan."

Vincent was fascinated by health care, a business in which he saw certain fertile similarities to semiconductors. He realized a few things others overlooked: first, advanced technology brought with it an increasingly high rate of change; second, there were inherent differences between drugs and diagnostics. These simple insights had significant implications. "One of the reasons that I had a great interest in joining the health-care industry . . . was my perception that that health care was an unusual industry back in the seventies," he said in an interview a decade later.

> You had a huge industry growing very rapidly, largely based on science and technology. I noticed that some of its managements had difficulty dealing with the high rate of change in the environment, which usually goes along with high growth rates and high technology. My somewhat simplistic intuition about why this was the case had to do with the fact that the industry came from a pharmaceutical base. Pharmaceutical products take longer to develop, and this has major implications for management. . . . It [diagnostics] was fundamentally a different business from the drug business.[3]

With the hiring of Vincent, preparation for battle began. Vincent summoned shock troops who had served with him in the chip wars, notably Jack Schuler, who had worked under him in the Far East and Europe, and who was unhappily stewing as a strategic planner in Texas. Schuler would eventually succeed Vincent and go on to become president of Abbott under Schoellhorn, who, in 1981, took over as chairman. Vincent also hired George Rathmann to run research and development. Vincent was a convert to the strategic gospel of market share; research was just a means to an end, the first step in a campaign to dominate the business. "When Vincent started," said a former Abbott manager, "part of his theory was that there's no IBM, Xerox or General Motors in diagnostics. Everybody had 1 or 2 percent of the pie—or less. Vincent believed that when you get to about 10 percent you became a dominant force in the market. His goal was always to be the market share leader. It was the same theory the Boston Consulting Group was selling: the market share leader controls the market. Go out there and get as big a share as you can."

Vincent and the Texas Instruments gang moved swiftly. Military metaphors abounded: total victory; attacking new markets; take no prisoners. New employees were "indoctrinated" in the first few weeks—a significant factor, for in three years, 80 percent of the division had arrived post-Vincent. Diagnostics got its own building. Vincent cultivated the fact that this division was *different,* even from the rest of the corporation. Diagnostics flaunted the fact that it had a quarter the number of secretaries per person that the rest of the company had. Salespeople had to operate on half the budget of their competitors. Available cash—and there was a lot more of it in total than a company like Genetic Systems could generate—went to hiring scientists and salespeople. "Even if you hire a scientist or salesman who's no good at all, they might do something by accident that adds to sales," explained Schuler. "Secretaries and accountants are part of overhead."[4] To complaints came the reply: Your time will come. "Vincent didn't have to fire anyone," said Philip Whitcome, who went on to become head of strategic planning at Amgen and subsequently chief executive of Neurogen, a Blech brothers startup focusing on the neurosciences. "If you weren't as dedicated as he was, forget it, he would drive you crazy."[5]

Abbott Diagnostics did not overflow with products: a few radioactive tests, a nicely profitable hepatitis test. Vincent poured cash into new immunoassays using polyclonal antibodies, an area that Syntex's Syva unit, a newcomer to diagnostics, had profitably pioneered. Schuler's marketing team fanned out overseas, setting up country organizations. No scrambling, local, third-party, fly-by-night distributors with suitcases full of deals—and competing products—careening around Brazzaville or Buenos Aires in rusty Studebakers. Schuler wanted Abbott representatives with business cards, suits, and regular hours. This took time and money, but it allowed Abbott to control its products, gave it a much larger sales base, and kept foreign competitors struggling at home, instead of rooting around in Abbott's backyard. Schuler believed in deep roots. "Medical markets are very conservative," he said. "Customers have to relate to life and death on a regular basis. They've seen a lot of disappointments, particularly outside the United States. If they have a reliable supplier they can count on, they stay. It's like the IBM security blanket. We would do things that you wouldn't usually do, like twenty-four-hour service guarantees, just to prove we were there to stay. It cost us, but once we got in, the thirty or forty others had a tough time dislodging us."

Markets were changing. Abbott poured money into antibodies and automated equipment. As the technology made testing cheaper and easier, it moved from central labs to hospitals to doctors' offices to homes. Semiconductor makers had seen this once before: as chips shrank and exploded in power, computers invaded offices, homes, and briefcases. Most companies sold either esoteric hardware such as radiation detectors and huge, whirling, clanking analyzers, or supplies called reagents that allowed the analyzers to perform specific tests. You could think of it as hardware or software. Abbott had been a reagent supplier and had depended on outside suppliers for instruments, but that made servicing very difficult. So, early on, Abbott began buying gamma counters required for its early radioactive tests and reselling them in a package with reagents—hardware *and* software. Its first attempts at building its own automated instruments were disastrous. "We stubbed our toe," said Schuler. "But we went to school, and at the end of that time, we knew how to make reliable instruments at low cost." Once a machine was placed, Abbott made money selling supplies, mostly reagents. Customers found it more convenient to deal with one supplier, and Abbott could offer more flexible pricing. And because Abbott was not designing either instruments or reagents to work on other company's products, its engineers could push performance, maximizing speed and ease of use, minimizing size and cost—and freezing everyone else out.

The master stroke was yet to come. It stemmed more from luck than analytical foresight. "These guys who write the history and think it's all great visions, that's a lot of baloney," said Schuler. "Sometimes you just fall into this." One day in 1978, Abbott found that an overly sanguine view of sales had left them with a warehouse full of unsold instruments. "We looked around and said, 'How the hell did this happen?' One guy said, 'Look, rather than have all that stuff sitting around in inventory, why don't we try a trial program? Go to our hospital customers and tell them if you run a certain number of tests a month, we'll give you a free analyzer.' The goof-up became a raving success." By giving away analyzers, Abbott forced its way into labs it had never been able to enter. That, in turn, pumped up reagent sales. That cash went into research, which resulted in new tests, new analyzers, more sales. Meanwhile, the rest of the industry scrambled to adjust. Instrument companies tried to buy reagent companies; reagent companies struck quick deals with instrument companies.

By the early 1980s Abbott was using its superior firepower—a combination of money, salesmen, and technology—to storm lucrative markets. For a big company, it moved quickly. In 1979, Abbott came out with its Quantum, an automated analyzer aimed at the large clinical chemistry market. Two years later, while Genetic Systems was signing

The Conservative Reaction

its sexually transmitted disease test deal with Syva, Abbott led the field with a 15 percent share of the antibody-based clinical chemistry tests. Right behind came Syva with 14 percent, followed by Johnson & Johnson's Ortho Diagnostics division with 7 percent, Baxter Travenol with 5 percent, and Warner Lambert with 4 percent; 140 other companies shared the rest. In 1982 it introduced the TDx, a second, even smaller automated machine. The TDx combined an immunoassay with an ingenious fluorescent marking system and a small computer—easy to use, easy to read. Its purpose was to monitor the levels of drugs—antibiotics, chemotherapeutic agents, and cardiovascular drugs—in the blood of hospital patients. At first, Abbott had developed only a handful of tests to run on the TDx, and sales mounted slowly. But inexorably, the company added to the menu, one a month in 1982 and 1983, all the time giving the breadbox-sized machine away, building the base.

Syva, which had pioneered the business, continued to dominate the market. Syva had a good analyzer of its own, a broad range of tests, and the largest installed base of machines. Why get nervous? Syntex had a record of strong R&D and besides, who was Abbott anyway? In a fast-growing market like this, there should be room for many competitors. That's the way it had always been in diagnostics. Besides, Syva had the rights to a whole new product line from Genetic Systems. And beyond that, they had a world-class scientist like George Todaro working on cancer diagnostics and therapeutics. By the end of 1983, Syva may have felt the pressure, but it was doing just fine.

For the biotechnology startup, Abbott did more than just compete: it changed the way the game was played; it made the game frightfully expensive to play. You could not make big money unless you developed labor-saving automated machines that could automatically grind out results. Then you had to *give* it away. You needed someone to take reagent orders, take customers to lunch, and sell. You needed a marketing team: minimum fifty people, making at bottom $40,000 a year. That's $2 million just in salaries, not including drinks, dinner, travel, and Christmas gifts—say another $30,000 to $50,000 per person—and that's not including a service department, manufacturing, and spare parts.

Abbott made strategic planning at a company like Genetic Systems, with its dream of leveraging diagnostics into therapeutics, an exercise in pyramiding contingencies. We will become a big company *if* we make enough in diagnostics to fund therapeutics. But we can hit it big in diagnostics only *if* we can make enough money off the antibodies we

119

license out, and the research deals, to fund more products that we can own outright—not only finance research and development internally, but develop our own manufacturing and marketing teams. If, if, if; with each if, the odds mounted. More attention came to focus on ways not of making money, but of raising money. Playing the game required cash. Abbott's Schuler called it "a Darwinistic process" taking over diagnostics, in which money, size, and market share became increasingly important. Abbott, of course, had it all. "If you look at other high-technology industries that developed after World War II, when trade barriers were removed, one would predict that there will be one leader, then maybe another company that will be one-half the size of that leader, and then a third company that'll be one-half the size of that," he said at a Dean Witter diagnostics forum in 1983."[6]

At that same meeting, Schuler also warned: "I think a lot of signs indicate we're moving into the consolidation phase."[7] Two months later, just as the market was peaking, the federal government was poised to crack down on escalating health-care costs. Medicare was nearly bankrupt. Social Security looked as creaky as a pensioner. In October, Congress mandated a new Medicare payment scheme which, in effect, punished hospitals for running costs up. This was a major sea change, with repercussions that would take years to sort out. One of the first results was that hospital purchasing, save for necessities, plummeted. Unsure of the long-term effects of the new system, hospital administrators tried to make do with what they had or turned toward integrated suppliers who could provide everything at a low price. Abbott thrived. Not only could it supply both machines and test kits, but it *gave* the machines away, allowing hospitals to avoid an item in the capital budget. And Abbott convinced doctors and hospitals that efficiently performed testing could get patients out of the hospital faster, saving money. Abbott gobbled up huge chunks of the market while its major competitors reeled. At the same time, the dollar was gaining strength against foreign currencies, shrinking the value of overseas sales when translated back to dollars. And approval time for new products in Japan, Germany, France, and Latin America was getting longer. This hurt everyone with overseas sales, but particularly the companies already hurting. Many companies, large and small, began to bail out of diagnostics; others stayed and suffered; a few found refuge in specialized niches.

One of the big losers was Syva. Syva had built its franchise on immunoassays, sold by its hundred-person diagnostic marketing team. In 1983 it broke the $100 million sales barrier, due in large measure to the success of its drug-monitoring tests; although it was beginning to sell Genetic Systems' Microtrak tests for chlamydia and herpes, those products contributed little to profits. Then, a few months into 1984 the bottom dropped out. Abbott had been continually adding new tests to its own drug-monitoring system, the TDx, and steadily gaining ground. Under pressure from Abbott, squeezed by Medicare, Syva took a beating. Indeed, it lost $19 million in 1984 on sales of only $92 million.

The ripples washed up in Seattle. Genetic Systems, of course, was to receive royalties on Microtrak. To be successful, the tests had to be peddled hard, and customers had to be educated, from physicians who ordered the tests and prepared the samples to lab technicians who actually performed them. Syva's sales force was used to selling to clinical chemists in labs, not microbiologists, and not doctors—a problem. So too were the big, bulky fluorescent microscopes that the test required. Not every lab owned one, and to buy one meant hitting the closely watched capital budget. Syva also discovered that it was not alone in the market. Abbott soon introduced a chlamydia test called Chlamydiazyme, which worked in four hours or less; Johnson & Johnson's Ortho Diagnostics bought a small company called Immulok, which had some sexual disease tests; and DuPont was readying a gonorrhea test called Gonocheck. Meanwhile, Syva struggled to stanch the losses in its big drug-monitoring market. Who had the time, or the resources, to worry about a new market that needed nurturing when the ship was sinking? Nowinski raged but Microtrak languished—particularly after some early tests had to be returned. For Genetic Systems, 5 percent of a market that just refused to develop . . . well, it did not amount to much.

CHAPTER 13

Just a Few Technical Problems

I F BIOTECHNOLOGY found companies like Abbott surprisingly effective competitors, part of the explanation could be traced to the technology itself. By 1983, around the time that many of the more optimistic forecasters had predicted an onslaught of biotech products, researchers were discovering that the biological landscape was not as easy to conquer as they had expected. Time after time, complexities cropped up—some large, but most, irritatingly small—that created delays and inflated expenses. No single problem was, by itself, large enough to bring down the edifice of immunotherapy that had been erected; no single failure could fully defeat the power and allure of the magic bullet. An interferon might fizzle, but there were always other immunomodulators waiting to be researched and developed. Indeed, by 1983 interferon had slipped into the background, replaced in the press and on Wall Street by the promise of monoclonal antibodies. But once again, even as the drumbeats for monoclonals were beginning, researchers were discovering a series of barriers, of complexities and subtleties, blocking their way.

The area of greatest difficulty developed around the disease with the greatest promise—and the greatest hype: cancer. According to theory, tumor cells were studded with distinctive antigens that said, to swarms of antibodies that could read them, *cancer*. The body would then naturally generate antibodies that lock onto those antigens when a cell turns cancerous; then they are recognized by macrophages, or white blood cells, which descend upon and ingest them. Viewed this way, the onset of cancer could be seen as an immune problem; one only developed cancer when the immune system weakened. Like other cancer theories, like the viral or the oncogene theory, this had a pleasant, all-encompassing ring to it. Moreover, it tapped deep public sentiments. Like interferon, antibody therapy seemed natural, nontoxic, logical; and it offered the possibility of prevention.

Nothing fundamental threatened the simple, enticing brushstrokes of antibody therapy. The problem, however, had to do with time and its business corollary, money. Time was slipping by as companies wrestled to achieve consistently positive results on real people with real tumors. The passage of time did not matter all that much to academic laboratories or even to the drug companies, both of which had stable sources of financing; indeed, to the drug companies, time was their greatest ally. The biotechnology companies, on the other hand, were connected by a financial umbilical cord to Wall Street. Technical problems meant delays which meant escalating burn rates and larger annual losses. It did not matter that these were technical problems that would be solved eventually. By raising money on Wall Street, companies like Genetic Systems had gambled that the translation of theory to practice would be a simple and easy task. It was not.

What were the technical problems when it came to anticancer antibodies? First, the *naked* magic bullet theory—the notion that antibodies alone could eliminate tumors—proved disappointing in practice. Except in a few cases where high doses of antibodies were temporarily successful on a few lymphomas, it just did not work. And fleeting success in mice, such as Nowinski had had, did not prepare anyone for the difficulties of human cancers. Tumor cells in humans proved extremely sly: not all cells from the same tumor displayed the same antigen; and, like the chlamydia bacteria, although with less consistency, tumor cells seemed to alter their antigens as the disease progressed, particularly at that most dangerous moment when the tumor metastasized, migrating throughout the body. Some cancer cells even pulled their antigens in, like soldiers ducking into a foxhole against incoming shells: no antigen in sight, no macrophages, no danger. And, in the case of solid tumors, particularly those prevalent and intractable cancers of the breast, lung, or colon, one could only reach the surface cells, not the mass of unhappiness within.

It proved difficult even to find antigens useful enough to make an *in vitro* test work. For one thing, many cancers do not shed their antigens into blood, lymph, or urine. More problematically, many tumor markers, or so-called tumor-associated antigens, existed in the healthy as well as the sick, and could not be used to distinguish between one kind of cancer or another. Thus, the small handful of common markers that were known could be used only to monitor the progress of the disease, not to provide early detection. For those antigens, at least, there seemed to be no magic bullet.

There was a bright side. Cancer cells did have antigens, which could be targeted by the correct antibody, or more accurately, by the correct mixture, or cocktail, of antibodies. What if you combined the power of the antibody, its targeting, with the power of a chemotherapeutic drug or a chunk of radiation? This left the promoters of natural therapies behind—radiation or chemotherapeutic drugs were anathema to them—but provided a wonderful elaboration on the magic bullet metaphor. Step up the firepower and one could graduate from bullets to guided missiles, loaded down with deadly payloads that could destroy upon impact not only the cell itself but also surrounding cells. And early work provided enough hope to generate a flood of articles along this line. A series of tests on patients with severe solid tumors, one weighing fifteen pounds, by Dr. Stanley Order at Johns Hopkins University, used antibodies provided by Hybritech linked to a radioisotope. Dr. Order coaxed a dramatic decrease in tumor size, in one case large enough to surgically remove it. The drumbeats began. "The most exciting adventure in the annals of modern medicine: The coming revolution in cancer therapy," declared the cover of a 1984 book called— well, it really was inevitable—*Magic Bullets.*[1]

But again, while anyone could announce a revolution, it was more difficult to actually kill the king, or even overthrow traditional cancer treatments. Order's results were dramatic, but they were not consistent. Sometimes it worked, often it did not. The same old problems cropped up, not only for Order but for Nowinski, Todaro, and every other antibody researcher working with cancer. One had to find antibodies with greater specificity to particular kinds of cancers; the more antigens on the surface of the tumor cell the better. But you had to be careful; one did not want those same antigens on healthy cells—the dreaded, if disconcertingly common, "cross-reactivity," which would, like friendly fire, destroy healthy as well as tumor cells. Drugs linked to antibodies that worked superbly in culture would often run amok in the body, attacking vital cells of the immune system which shared antigens with the tumors. One had to make sure that the antibody bound tightly enough with an antigen so it did not fall off or was not somehow deactivated by the tumor cell. One had to make sure that cells did not withdraw their antigens on repeated administration. One had to make sure that the body did not flush them out through the kidneys or collect them in the liver.

And there were mice problems. Although mice antibodies look a lot like human antibodies, they do not match *exactly*. The body's immune system knows better, and sometimes, when physicians pumped a patient full of mouse antibodies, they triggered an allergic reaction; patients sometimes went into shock and died. It was not all that different from the allergic reactions that often developed after pumping someone full of interferon. Even worse, the immune system treated the mouse antibodies as invaders, as foreign antigens, and did its best to deactivate them. The body would develop a resistance to the mouse antigens, eliminating any benefit of repeated administrations. Finally, nobody was about to suggest using humans, like mice, as farms to grow human antibodies.

In a sense, solving those essentially biological problems was the easy part. New antibodies were being isolated all the time. Researchers were getting a better idea of which antigens appeared on which tumors and when. And a sort of hybrid antibody, half human, half mouse, called a chimeric antibody, seemed to promise a dance around the rejection problem. But lurking out there was another pitfall whose solution fell more into the realm of chemistry than biology: how to link the drug or the toxin or the isotope to the antibody. It sounds absurdly simple. And, in fact, many ignored it, choosing instead to search through endless clones for an antibody so perfect that it would prove effective with even a sloppy linker between monoclonal and drug. And yet, for those listening carefully, there were discouraging murmurs. In that same Dean Witter diagnostics conference that Jack Schuler spoke at, Michael Wall, the chairman of Centocor, talked about antibodies as therapeutic tools. "We do not in our therapy studies, link the antibody to a drug, a toxin," he said, adding, almost as an aside. "The reason . . . is because we don't know much about conjugating."[2] Wall was hardly alone. Conjugating was indeed a barrier as formidable as antigenic heterogeneity but far less apparent. Conjugating difficulties were, as one executive said, "a dirty little secret."

Conjugating sounds dull. But think of the marriage bed: let us conjugate. The purpose is to link, or marry, an antibody molecule with a number of smaller drug molecules. This sounds simple, but the reality was extremely tricky.

Left alone, antibodies are exquisite marriages of form and function. As a weightlifter bends and strains to pick up a barbell, so antibodies, to molecular grunting and groaning, assume new shapes to adjust to the weight of the drug molecules loaded atop them. Often, the antibody will contort enough to block its active sites, diminishing its ability to bind with an antigen. Even worse, the drug molecule may moor itself right in the active region. In all these cases, piggybacking a drug onto an antibody reduces its ability to lock onto an antigen—in the parlance, reduces its affinity for the antigen—which means it just does not work very well, if at all.

Indeed, when scientists mixed the drug and linker with antibodies, far fewer linkages occurred than the theory would indicate. Conventionally, drugs are linked to antibodies through a handful of key building-block amino acids that offer handholds in the form of chemical bonds. A popular one is called lysine. The location and the number of lysine molecules, called residues, vary from antibody to antibody; thus, some antibodies have far more linkage points than others. Moreover, the linker does not care where it attaches, whether to the lysine residue on the trunk of the antibody, far from harm's way, or to the residue adjacent to the active site, in which case it may block the site. The probability of a successful linkage thus depends on the antibody and its arrangement of lysine residues. If, hypothetically, an antibody has two residues, one near the active site and the other far up the trunk, there's a 50 percent chance that the drug molecule will bind near an active site. That, in turn, will reduce the affinity—a rough measure of effectiveness—by 50 percent. As one monoclonal researcher said, "It's one thing to have a marriage and another to have a *happy* marriage."

These sorts of linker problems magnified the difficulty of finding an effective antibody against tumor cells. Not only did you have to find an antigen that hit a high percentage of similar cancer cells, *but* you had to play the mouse game until you found an antibody that had a strong affinity for that antigen. *Then* you had to hope that your linker did not diminish, or destroy, its effectiveness.

Problems had to be solved one at a time. One possible answer came from a physician and researcher, Dr. Thomas McKearn. McKearn touches our story at several points. As a young pathologist in the mid-1970s, he had tried to get cells to produce pure homogenous lots of antibodies. Kohler and Milstein beat him to it, but he was one of the chosen few to receive their mail-order hybridoma. A few years later, in

Just a Few Technical Problems

1980, while at the University of Pennsylvania, McKearn and two colleagues edited an influential volume called, appropriately, *Monoclonal Antibodies: A New Dimension in Biological Analyses.* Bob Nowinski wrote a paper, along with just about every other big name in the field. The volume became such a popular work that Plenum Press, a scientific publishing house, published a second volume, *Monoclonal Antibodies and Functional Cell Lines,* in 1984. In 1980, with the first volume just out, Robert Johnston, the Princeton venture capitalist, tried to convince McKearn to put together a monoclonal company. McKearn turned him down. Johnston then recruited Nowinski, until, of course, the Blechs beat him to the deal. In 1982, Johnston turned back to McKearn. Would he reconsider? This time McKearn agreed.

Cytogen under McKearn was not Genetic Systems East. The differences are instructive in the ways that business and technology interact. Nowinski emphasized finding the *perfect* monoclonal and got mice up to his ears. McKearn, by the early 1980s, was convinced that commercializing monoclonals would pivot on the way drugs and antibodies were conjugated, that is, on linkers. After he signed on, McKearn sat back and thought about where the company should go. For five months he and a former Penn colleague, John Rodwell, set up shop in a single office with a desk, two chairs, and a blackboard. "I don't accept the idea of research as a random process, closing your eyes and slinging darts," said McKearn. "I don't think you can know four steps ahead, but you should know where you're heading. The process, step by step, is predictable. We sat and thought, and challenged our assumptions, and we came up with a plan. The two major concepts of it were site-specific covalent modification and functionally active linkers."[3] In other words: Make sure the linker attaches to an optimal site on the antibody, and design those linkers so they can hold or release the drug, depending on the situation.

McKearn and Rodwell did not believe in the magic, undiscovered antibody. Useful antibodies had been found, they thought, including some at NIH that were free for the asking. As a result, Cytogen developed differently than Genetic Systems and Hybritech. Linkers were a chemistry problem, rather than a biological one. Linkers did not occupy as central a role in *in vitro* diagnostics, the kind of tests Genetic Systems was developing, as they did in therapeutics; thus the urge to use diagnostics as a running start for therapeutics, was not as strong. Indeed, after an early flirtation with *in vitro* diagnostics, Cytogen abandoned the business. Unfortunately, despite the logic of McKearn's approach, Cytogen never received the attention of Genetic Systems. Linkers were more esoteric than talk of magic bullets; they sounded mundane, like talk of plumbing instead of architecture. Many companies downplayed the importance of linkers. Even many who understood argued either that Cytogen would end up selling its linker technology to others who, in turn, would make finished products and keep the bulk of the profits, or that once the drug companies, with their legions of organic chemists got going, a hundred ways to link payload to antibody would appear. Nonetheless, Cytogen labored on. And in the summer of 1983 it began to make the preparations for its own initial public offering.

The timing proved to be unfortunate. By midsummer enthusiasm for biotech was cresting. This had less to do with commercial perils or technological drawbacks and more to do with the kind of generic collapse that follows every speculative bubble, from tulips to high-tech new issues. Too many technologically based companies—and this included a veritable flood of small computer as well as biotechnology firms—had gone public at dizzying prices. Bad money drove out good. The frenzy for technology fed on itself; long-term values had been lost in the stampede for quick trading profits; investors, including the eager amateur and short-term trader, poured their money into the market not in search of solid, long-term gains but with the blithe confidence that the market would continue to drive every new issue upwards. Inevitably, late in the summer, came the break. The market awoke to the realization that many of these newly created companies were junk. Prices collapsed; the offering window slammed shut. One of those companies caught without financing was Cytogen. The company had tried to go public, but by the time it had prepared the papers—an ironic counterpoint to the legal delay that allowed Nowinski to defect—the moment had passed. Cytogen would have to wait three more years.

Biotechnology, resting atop its thin base of money, arguably suffered from the collapse more than any other identifiable industry group. Trading in biotechnology stocks dried up and prices fell as investors, particularly the large institutions, fled to the safety of the blue chips. In October 1983, Genetic Systems took a particular beating after Nina Siegler, Paine Webber's biotechnology analyst and a supporter of the company, left the brokerage house. Rumors immediately began to circulate. "Strong selling occurred to Genetic Systems stock over the last week instilled by primarily two events," said a report from another brokerage firm, Piper, Jaffrey & Hopwood:

First Paine Webber lost two biotechnology analysts and indicated to its brokers that it would no longer have coverage on a number of biotech stocks. Genetic Systems was one of the biotech stocks mentioned. Second, there has been some concern that a recent 144 stock sale [blocks of stock distributed before a public offering, such as shares held by the Blechs or Nowinski] may have been insiders selling off the stocks.[4]

Actually, neither of these events seem to have taken place. Paine-Webber replaced Siegler, who went to Prudential-Bache, with Linda Miller, who in turn picked up the coverage. And no major inside holdings were sold. Nonetheless, rumors are symptoms of nervousness. Wall Street was looking for a justifiable reason, short of admitting to having made a mistake on fundamentals about Genetic Systems, to dump its stock.

Still Genetic Systems was not alone; the damage was general. Companies were left waving promises to investors who were running the other way. Some biotechnology executives, as if innocent of the essential amorality of Wall Street, talked like jilted lovers. It wasn't fair. Last year, you said you loved me. Last year you believed my promises. Now you drop me cold for . . . some appliance company.

Others prepared for hard times. Cetus hired Biogen's president, Robert Fildes, a former pharmaceutical executive, hoping he could make some sense of its tangled operations. Biogen under Nobel laureate Walter Gilbert staggered under large losses. Other companies simply disappeared from sight. Only Genentech took it with aplomb. Swanson told analysts: Genentech's first product, a brain hormone called somatostatin, was really a demonstration of the technology; and its second, human insulin, which has just been approved, has not gotten off all that quickly. But we proved we could do it, and now we have some winners: alpha interferon, human growth hormone, and a protein that dissolves clots that cause heart-attacks, tissue plasminogen activator, or t-PA. "We'll be a billion-dollar company by 1990," he declared over and over again. "One billion by 1990" T-shirts soon appeared at Genentech Ho-Ho's. An analyst recounted how he had once sat with Swanson at a lunch and asked him, "Do you really believe that?" Swanson, without missing a beat, whispered, "I really said the 1990s." The market was not privy to that admission. Genentech's stock never fell much below thirty-five, its offering price, the price at which it had gone public.

The monoclonal companies had no interferon, no t-PA, no Bob Swanson, although both Genetic Systems and Hybritech did have Friday beer bashes. What they could hold out was the promise of monoclonals against cancer. But they had painted themselves into a corner; when you talked magic bullets, pregnancy tests just did not cut it. When you painted visions of revolutions in health care, chlamydia hardly made the grade. The name of the game, at least on Wall Street, continued to be cancer. It was a trifle late to start talking about the complexity of the disease and the scientific problems. The unfortunate truth was that investors who had bought the stock more out of speculative enthusiasm than understanding the first time around were not going to try to figure it out now. However, one could hardly escape the reality: there were still difficulties to be overcome. Just a few. Technical. Problems.

CHAPTER 14

On the Far Side of the Bubble

READING OLD REPORTS on Genetic Systems creates a curious sense of disequilibrium. For the first three years or so, until the second offering in 1983, analysts seemed to have Genetic Systems cold. Deals are signed, projects are launched, and money is raised. Projections of revenues, based on these contracts, offerings, interest income, are not wildly awry; estimates of losses—expected losses—hit home within a few cents.

Then, in 1984, it begins to break down. Analysts had pointed to 1984 as the year Genetic Systems would begin ringing up product sales, perhaps earnings; the boom in the stock price of the prior year, including the offering, had been based upon that expectation. The projections, alas, even among those not ready to foresee profits, were quite far off the mark. It was difficult, at first, to locate the cause. The FDA, the usual scapegoat, up to that point had approved the new tests with reasonable dispatch.[1] Researchers at Genetic Systems had, under Nowinski's whip, successfully cranked out antibodies for the sexually transmitted disease tests and for the respiratory infection tests against bacteria such as pseudomonas; not everything went right, but Nowinski and his team were respected enough to have written a general overview of monoclonals as diagnostic tools against infectious disease in *Science* in 1983.[2] The Blechs on Wall Street had, despite a crumbling stock market, generated plenty of cash for Nowinski to spend.

Who then? Or what? The distress could be traced to a series of interlocking factors. First, with the market bubble having burst, new financing was hard to come by and burn rates were accelerating. Second, the environment for health care products was suddenly concerned about cost; no longer would hospitals automatically buy more expensive technology if it promised only marginal improvements, say speed or just a bit more ease of use, over inexpensive, well-established, if slower, methods. A clear advantage on cost and efficacy now had to be proven. Third, a powerful resurgent foe like Abbott made the game more expensive to play. And finally, add to this the dark side of serendipity, a kind of Murphy's law where misfortune, rather than miracles, rules; the kind of technical problems that drug companies regularly factor into their product schedules, their budgets and financing, but which the wilder promoters of biotechnology tended to view as needless pessimism.

Finally, there were the internal failures of judgment. Genetic Systems underestimated the rigors of diagnostics and of the resistance to new tests sold by new companies and overestimated the allure of the new technology. All this would have made little difference had not Genetic Systems raised expectations in order to sell investors. Wall Street may have a selective memory, but it does have one: once those expectations soared, fed by investment meetings, press releases, analysts' reports, and the tinder of idolatrous stories in the press, failure to fulfill them created its own unfortunate consequences.

The grinding reality soon made itself felt. The Syva tests produced so little royalty that the company did not even break them out separately on its balance sheet as sales. The chlamydia test was lucky to ever amount to $10 million; in 1986 *all* the Microtrak tests sold about $10 million worth. The failure, in turn, placed a heavy burden upon the respiratory tests. Nowinski had predicted that Genetic Systems would have a full panel of tests—for streptococcus, staphylococcus, Legionnaires' disease, and three forms of pneumonia—by late 1984. He was, however, wildly optimistic, and for the most part, the analysts followed him. Despite the fall in stock prices, the analysts still put their faith in Nowinski's scientific prowess; and beyond him, Todaro and Oncogen.

Where were the problems? Well, even the optimist should have realized that clinical trials would absorb at least three months, followed by some time at the FDA awaiting approval, followed, in turn, by FDA certification of the manufacturing facility. Analysts, in particular, tended to assume that sales would begin to mount the moment FDA approval

On the Far Side of the Bubble

arrived. They rarely did. Wall Street then was only beginning to realize the intricacies of manufacturing biologicals, and it was widely viewed as something that could be taken care of with a few dollars. Nonetheless, significant fundamental problems remained. As one analyst commented, almost as an aside in 1983:

> Furthermore, the quantities of antibody which would be necessary for commercialization will require an effective scale-up plan. At the present time, there are significant differences of opinion about the most likely scale-up processes. Some speak of new ways of mass culture of mammalian cells (bypassing the mouse system which is efficient for small amounts of antibody but overwhelmed by bulk product demands); others believe that fermentation provides the answer, requiring scientists to engineer the proper genetics into organisms such as bacteria or yeast, in essence making recombinant monoclonal antibodies. At this point there appears to be no consensus.[3]

That fact had very little effect on the outside world. Engineers, particularly bioprocess engineers, worked among tanks, vats, pipes, and dials; they were considered mere applications guys, far less important than the bench scientists. This proved to be a dangerous prejudice both internally and externally. Actually, biotechnology was, above all else, a manufacturing breakthrough: a means of manipulating the internal works of a cell so that it can produce biological substances on demand. To make it work required a sensitivity to manufacturing at the earliest, genetic engineering stage.[4]

Few companies had that sensitivity. The Legionnaires' test was, after all, Genetic Systems' first fully made commercial product. The company had made monoclonals in research quantities, but that was not the same as producing it in commercial volumes; with larger quantities, the parameters and uncertainties multiplied. To gear up, the company had to hire and train new people. All that took time under the best of circumstances. The FDA approved the first test, for Legionnaires' disease, in August 1984, but marketing did not begin for nine months— in April 1985. Much of that time was absorbed in getting manufacturing straightened out and the facility approved by the FDA.

Even then the tests did not sell all that well. Like the Microtrak tests, the Legionnaires' test required a cell culture and a fluorescent microscope. As Nowinski promised, the Legionnaires' test provided an edge over older tests which used as many as six antisera, or reagents, and produced bunches of false positives. But Legionnaires' offered an example of the mass media fallacy—that is, just because it's news it must be important. The microbe that killed thirty-four American Legion con-

133

ventioneers in 1976—*Legionella pneumophilia*—was indeed famous. But the market for Legionnaires' testing was estimated at only $200,000 a year (although, as one report said, "growing at a healthy compound rate of 15 percent a year"). And the company, which still lacked a sales force in the field, tried to sell it by telephone, a cheap, but not very effective marketing technique. Moreover, 5 percent of the sales that did trickle in were also diverted to the limited partnership. Thus, like Microtrak, the Legionnaires' test left but a dusting on the bottom line.

Still, it was nothing to get overly concerned about. Genetic Systems had other respiratory tests lined up behind Legionnaires', as well as $37 million in the bank. That sounds like a lot. But in pharmaceuticals that's about a third of what it takes to get a single drug to the market; and even in diagnostics, cheaper by far, that cash would evaporate once a real marketing group was formed. Besides, what did a few months delay mean? Not much—until one remembers that Genetic Systems was only in diagnostics to help launch itself into therapeutics.

In 1984, Genetic Systems was like a high-performance engine that had developed a knock. It seems, at first, to be a very minor problem, and indeed, in a motor scooter, one might dismiss it. But leverage has two sides: one requires a little effort to construct a soaring edifice, the other a tiny flaw to tear it all down. Thus, the pyramid scheme, the perfect model of the leveraged construction; question its stability, remove a single brick, and down it tumbles. Leverage of whatever kind, by definition, requires a risky leap across the void. In 1984 the ability of Genetic Systems to make that leap began to look questionable.

In early 1984, Genetic Systems' stock fell to just over five dollars a share, not far from its book value—the value of its assets (including cash) minus its liabilities—of two dollars. No one would have cared much if not for those B warrants. Those warrants, a relic of the first 1981 offering, enabled holders to buy Genetic Systems stock at five dollars a share. With the stock hovering around five—and no one knowing how much further it might fall—there was no profit from taking advantage of the deal. At the end of March, three months or so before the deal expired, the company extended the deadline a full year to June 1985. That meant that instead of having an additional $16 million in cash in the bank—and over a million more in interest income reducing its losses—the company was left, for the moment at least, with nothing.

The vise slowly tightened. In 1983 the company lost $1.5 million, a manageable amount of money considering the expectations. For every dollar burned up in internally financed research and development, the company scared up two more in contract payments from the likes of Cutter or Daiichi (which it then spent). And the public offering put $21 million in the bank. At that rate, Genetic Systems could last twenty-five years before it burned up the last of its cash. This, however, provided a deceptive sense of assurance.

The Syva deal and the money from the respiratory partnership would both end in the fourth quarter, and the Cutter deal on developing an antibody to pseudomonas would end in 1985. The researchers, the technicians, and the secretaries working on those projects would have to be paid, of course, or laid off, a dangerous precedent for an entrepreneurial company—bad for morale, bad for public relations. The original idea, of course, was to replace revenue from research contracts with cash flow from products or from new contract jobs. Instead, while sales trickled in, new costs associated with commercialism mounted: manufacturing, a sales team, clinical trials, waltzing with the FDA. Genetic Systems needed a big product, a blockbuster, to get out of the hole.

As early as 1983, Bob Nowinski was talking about a machine. Abbott had its analyzers, Quantum and TDx, and Syva marketed its Emit system for drug monitoring. Hybritech, reeling from Operation Neutral—in whatever form that took—was developing its own analyzer, the Icon. The logic was implacable. Without a machine, the have-nots were at the mercy of the haves. Without a machine, Genetic Systems would face a future of squeezing out tiny profits no matter how well its antibodies worked. Without a machine, Genetic Systems would end up as a niche player, a reagent supplier, a slave to fortune and misfortune. That, at least, was the argument.

And so it went: certainly, the sexually transmitted tests and the respiratory tests were improvements—breakthroughs even. But they obviously did not sell themselves. Most infectious disease tests are ordered by physicians, from large clinical or hospital laboratories. And from the doctor's profit perspective, that money disappeared. Doctors were being squeezed on all sides—by insurers and patients and too many other doctors. Really selling those tests, Nowinski came to believe, required a dose of technology: an automated, accurate, fast, ten-to-fifteen-minute system that could be sold in one configuration to doctors and another to the labs—a machine that could make money for doctors.

Nowinski thought he had just the item. He had been working with a polymer chemist at the University of Washington who had developed a way of precisely attaching antibodies to a polymer, a sort of plastic, for use in an analyzer. And so in November 1983, Genetic Systems filed for a patent on the technique.[5] Building such a system, however, was beyond its capabilities; it required engineering resources and skills foreign to most biologists or to most biotechnology companies. These systems were packed with electronics, including sophisticated microelectronics, and optics; the complex software to make such a machine operate created enormous problems; and again, a system had to be manufactured efficiently, sold, and then serviced.

Here again, however, Nowinski, with his aggressive charm and scientific reputation, wooed an excellent partner: a California company called Applied Biosystems. Its name was apt: Applied Biosystems made machines that automated a variety of laborious laboratory processes. One of its machines synthesized, or assembled, strands of DNA; another built strings of amino acids which formed small proteins called peptides; a third broke down proteins into their amino acid constituents. In the lab they were generically known as gene machines. Applied Bio was a company that arose from the confluence of two technological streams: biology and electronics. Historically, this was a tremendously important blend; in actuality, Applied Bio kept a relatively low profile. It was the Levi Strauss of biotechnology, prospering by selling supplies to the gold miners.

Applied Bio was looking for new markets. There were only so many research laboratories to sell into, and diagnostics loomed as a natural future market. If Applied Bio could make automated systems such as DNA synthesizers, which used reagents and complex chemistries and which sat atop a lab bench, how difficult would it be to construct a test system, particularly if a partner like Genetic Systems was willing to contribute patentable technology, provide cash, and defray the risk? If it failed, Applied Bio could continue mining an already rich vein. Unlike Genetic Systems, Applied was already a profitable operation. It would not be balancing its current and future product lines upon a single machine.

The creation of deals like that with Applied Biosystems has to be seen as part of the evolving politics within Genetic Systems which were increasingly dominated by Bob Nowinski. It was Nowinski's nature to be aggressive, and he was clearly the dominant personality within Genetic Systems. Glavin, the president and chief operating officer, was a pleasant, likable soul, modest and flexible—perhaps to a fault. He did not come across as a whiz kid, as so many at Genetic Systems did, but as a solid, sensible manager. He admitted to few pretensions about his

On the Far Side of the Bubble

scientific insights, deferring in interviews to the brilliance of Nowinski or Todaro, who in turn thought him rather thick. He seemed to take pride in his ability to mediate among difficult personalities; he seemed, even years later, genuinely impressed by Nowinski's energy and intelligence. In investor meetings, to analysts, and to the press, he and Nowinski presented themselves as two parts of a smoothly functioning team, conforming to public expectations of a science-based company: Glavin as the business brains, Nowinski as the science. That certainly was reflected in their titles: Glavin, the president and chief executive officer, then chairman of the board; Nowinski, executive vice-president and scientific director, then chief executive officer.

Titles, however, failed to reflect the political realities. Glavin lacked the power of Swanson at Genentech, Howard Greene at Hybritech, or Hubert Schoemaker at Centocor. Nowinski had the loyalty of the Blechs, and he spoke a language Todaro understood. Glavin was dispensable, Nowinski was not. Not that Glavin did not struggle; the battles, for the most part, took place behind closed doors. Even years later, he was too loyal to reveal his conflicts openly. Only the occasional sign of tension escaped to the outside world. Relationships with scientists, he told a reporter in 1983, "depend on how you approach them. I make the assumption that I don't know anything about science, so I never second-guess on it. . . . And frankly, I object if I hear a [scientist make a] very superficial or facile generalization about the *marketing* of a product." Nowinski, in turn, agreed. "We're both very strong-willed and competitive individuals and, when we disagree, we sort of bang it around. The most important element is mutual respect."[6]

Nonetheless, to analysts and money managers visiting Genetic Systems, the situation was clear. Many came away disturbed by Glavin's increasingly subordinate status. "It was not being run like a real company," said one money manager who visited in 1983. "Glavin was a puppet for Nowinski. He was president, but I spent three hours with him and couldn't figure out what he was talking about. It was clear Nowinski wanted to run everything."

Nowinski could be arrogant. At Fred Hutchinson and at Cytogen he had left behind a trail of simmering antipathy. At Cytogen, before jumping ship, he had clashed with others because of his belief that he knew what was best in such business disciplines as marketing. "His style is to emasculate you," said a colleague who worked with him before Genetic Systems. "He's always right, you're always wrong." Now his power, and his willfullness, steadily increased. The joke began to circulate: Bob thinks his nickname is Dr. Now. It's really Dr. No. Here comes Dr. No. Watch out for Dr. No. His aggressiveness also worked itself into an increasingly bold strategic plan. Rather than retrench as the market turned sour, Nowinski pressed restlessly onward, seeking a score that would break the company out of the cycle of dependency.

His opportunity came with the rise of acquired immune deficiency syndrome, AIDS. The first AIDS cases were identified in 1978 in New York and Port-au-Prince, Haiti. It spread steadily, infecting specific groups: Haitian boat people, intravenous drug users, hemophiliacs, and large numbers of male homosexuals. By November 1984, well over 6,000 victims had been diagnosed as suffering from AIDS—70 percent of them homosexuals—and almost 3,000 had died. Like herpes, it was incurable; unlike the sexually transmitted diseases, with the exception of untreated syphilis, it delivered a death sentence. No one was laughing. And yet AIDS research showed, for the first time publically, what molecular biology could do: after barely two years of research, two organizations announced, in the summer of 1983, the isolation of what they believed was a causative agent, a remarkable achievement unthinkable just a short time before. In France, the Pasteur Institute declared AIDS to be caused by a virus called LAV, or lymphadenopathy-associated virus. In America, Robert Gallo's lab at NCI called it HTLV-3, the third member of a family of retroviruses called human t-cell leukemia virus; these viruses were similar to the retroviruses so important in oncogene work.[7]

In terms of research, AIDS and cancer had much in common. Gallo's laboratory had earlier discovered both HTLV-1—a virus that seemed, finally, implicated in a human cancer, leukemia—and HTLV-2; and the AIDS virus seemed related to a virus that caused cancer in cats, the feline leukemia virus. This link between a virus and a cousin to cancer gave George Todaro a larger role. He, like many of his NCI colleagues, had drifted away from viruses. Now they began coming back.

Despite the unseemly squabbling between NCI and Pasteur over who discovered what and how and when, LAV and HTLV-3 appeared identical. NCI turned to more practical problems. AIDS could be transmitted through blood, and NCI needed a test that blood banks could use to screen infected blood. Gallo's laboratory, the epicenter of American AIDS research, finally chose five companies to produce a test based on its HTLV-3 virus: Abbott, Electro-Nucleonics, Litton Bionetics, a joint venture between DuPont and Biotech Research Laboratories, and one between Baxter Travenol and Genentech. Chiron and Centocor also announced that they were working on antibody tests that recognized discrete AIDS antigens.

On the Far Side of the Bubble

In early 1984, Nowinski leapt into the growing AIDS sweepstakes. Genetic Systems would produce a diagnostic based on the LAV virus from Pasteur, forming a joint venture with the institute's commercial arm. The Pasteur group had applied for a U.S. patent on LAV months before Gallo's NCI lab on HTLV-3. If Pasteur had won the suit, Genetic Systems, as the sole U.S. licenser, might well have been the only legal marketer. Was the AIDS test a blockbuster? The estimates looked solid: anywhere from $100 to $250 million. That would make it the largest diagnostic test of all. Unlike with chlamydia or herpes or Legionnaires' disease, blood banks would have to test for AIDS. Moreover, American Hospital Supply, with its large sales force, agreed to market it. Thus, for once, there was little risk that the market would turn out to be smaller than the numbers. Once a test appeared, the blood banks would quickly welcome it. The only issue was the up side: how big would it get, how profitable, and for how long?

On July 18, 1984, Glavin and Nowinski appeared at Cable, Howse & Ragen on Fifth Avenue Plaza, in the heart of downtown Seattle. The stock had not yet recovered, the warrants were adrift, and another limited partnership sales blitz loomed. It had again come time to send a message out. Investors needed to be reminded of the glorious future.

The pair talked to analyst Robert Kupor. Their discussions ranged widely, touching on the sexually transmitted disease tests ("the chlamydia test is the most successful," they reported to Kupor), the analyzer project ("the real payoff in diagnostics will come when diagnostics can be run on rapid, automated machines"), and the Cutter pseudomonas project ("marketing data suggest that the total market value for MABs against gram negative bacteria . . . could be $1 billion"). Mostly, however, they talked about AIDS: market size, patent speculation, LAV as compared to HTLV-3. AIDS and the tests bundled with it—hepatitis B and a viral cousin to HTLV-3, HTLV-1—would represent the company's most ambitious project yet, they said. As a result, they were pouring 20 to 25 percent of its research and development resources into the AIDS test, with the intention of introducing it between January and March of 1985.

The company [wrote Kupor to his clients] may decide to introduce the test in conjunction with an automated diagnostic instrument, which would be competitive with or superior to the Abbott Quantum that now dominates the markets. Genetic Systems feels that it has an advantage over competitors because Dr. Nowinski's own background (along with that of many of his colleagues at Genetic Systems) has been in AIDS-like viruses. It was largely because of this technical expertise that the Institute Pasteur Production chose Genetic Systems as its American collaborative partner.[8]

Only one false note rang out. Kupor told clients that he was increasing his estimate of losses from six cents a share in 1984 to a loss of eleven cents, an 80 percent increase. But, he added, he expected his 1985 estimate of a five-cents-a-share profit to be accurate. He urged accumulation of the stock.

CHAPTER 15

The Best Big Business in the World

I N THE YEARS following World War II, no major industry has been as steadily and increasingly profitable as the pharmaceutical industry—at least, no legal industry. That prosperity was fueled and protected by two fundamental underpinnings, one technological, the other legal. The first, of course, was mass screening, which made drug discovery a long and expensive, but ultimately productive, process—a strange blend of the industrial and the serendipitous. In latter days often poked fun at, mass screening in its heyday still produced scores of powerful, "miracle" pharmaceuticals. The second was the product patent, which gave companies long years of marketing exclusivity for proprietary products. The ethical drug industry grew up around those two solid, indeed increasingly symbiotic, supports. The difficulties and enormous expense of mass screening were used to justify the granting of exclusivity. And the large profits that came from patented products made the drug companies slow to look beyond screening as a research tool.

By the 1960s the dozen or so major U.S. drug companies were the prototypical Republicans of corporate America: large, rich, and very conservative. Companies with patented drugs could become cash machines, particularly as manufacturing facilities were paid off and the need for marketing well-known agents declined. Without competition, with no one questioning health care costs, they could set prices freely.

And because illness paid no heed to economic cycles, they made money in good times and bad. Thus, organizationally, drug companies tended to be very stable. Few new companies could afford to enter their ranks—only Syntex had managed to do so since World War II—and management turnover was very low. Short of a recall or a scandal, top managers would spend ten or fifteen years at the top; they might be long retired before their successor discovered a bare cupboard. Even then, products going off patent faced little competition. Trying to take business from another company's off patent drug was viewed as an ungentlemanly business.

Prosperity created a monolithic character to the industry. Firms belonging to the Pharmaceutical Manufacturers Association (PMA), a trade and lobbying group, formed an intimate club with its own unwritten rules. Unseemly competition was frowned upon. Criticism within the industry was muted. Hostile takeovers were taboo. Competition was pursued on technological grounds, with patented products. As we saw before, medicine generally took refuge from uncertainties in conservative behavior. The drug industry's major customers, physicians, and its regulator, the FDA, built up and shared a bureaucratic mentality: it was always better to be safe than sorry. And this familiarity bred a sense of security.

But by the 1970s there was trouble in this corporate paradise. Research productivity was falling, fewer drugs that opened up major new markets were being developed, and large numbers of similar products were crowding into popular therapeutic categories. The economics of pharmaceuticals began to change. Antibiotics, the largest drug category and the fuel for the glorious growth of the postwar years, were particularly exposed. By the mid-1970s there were few major microbial infections that one antibiotic or another could not handle. More products chased after the same business, old standbys began to lose their patents, and even price competition, albeit minor, appeared.

Still, the companies continued to make money, supplemented by nondrug acquisitions like cosmetics, candies, diagnostics, or medical equipment; markets were orderly and the situation seemed to be in hand. Then came two hammer blows. In 1983, Congress lowered the regulatory hurdles for companies wishing to sell a generic, or off-patent, product. The result: dozens of small, generic houses rushed in to undercut the giants. Soon after came hospital cost containment, which forced hospitals to create a market for generics where price was a factor.

Of course, companies with new, patented products had no need to worry. If patents had been important before, they were now critical as

The Best Big Business in the World

protection against the chaos of the competitive marketplace. But where were those patented products? The companies found themselves caught in a dilemma. On one hand, a reliance on screening would no longer do. On the other, the confident predictions of a new generation of biotechnology products as "revolutionary" as the antibiotics seemed to many companies either unrealistic or beyond their immediate reach. The companies knew they had powerful assets: money, marketing, clout at the FDA, and well-known brand names. Drug markets had always been conservative markets. They would move conservatively. They were, after all, still making immense profits. So, one by one, they began pouring money into new laboratories, hiring biologists and, in a few cases, licensing products like alpha interferon or human insulin. Research budgets crept to 8, 10, 12 percent of sales, higher even than in the semiconductor industry; many of the companies now began selling off those same nondrug units they had bought a decade or so earlier in order to free up cash for drug research. They were, in a sense, preparing to play both sides of the street: building a capability to do the kind off inductive, basic research molecular biology was famous for while funding traditional screening programs.

They were gambling—in their own peculiar fashion. If biotechnology was as revolutionary as advertised, they would face a serious threat. But history had taught them that the pharmaceutical markets were hard, exacting arenas. So if the new biology turned out to be a basic science in search of applications, they would just pick up the pieces when the revolution fizzled. By 1984 many drug company executives began to sense they had won their gamble.

The Bristol-Myers Corporation was, by any standards, a member in good standing of the pharmaceutical club. Its headquarters resided in a blank corporate tower at 345 Park Avenue, Manhattan. The building sat stolidly between booze and beatitude: to the north rose Miës van der Rohe's elegant Seagrams Building, to the south, St. Bartholomew's Church, with its Byzantine dome and garden. The Waldorf Astoria Hotel attracted limos and dignitaries a block away; across the islanded avenue stood ITT, Colgate-Palmolive and Lever Brothers.

In 1984, Bristol sold over $4 billion worth of products and earned almost $500 million—approximately five times the amount that Genetic Systems had been able to *raise* since 1980. Its profit margin swelled for the twelfth year in a row, the cost of products compared to sales fell for the third straight year, and it had $800 million in the bank and a mere $100 million or so in long-term debt. Bristol had increased its dividend to shareholders every year since 1972. And, even more re-markable, it had increased its earnings every year for the past quarter of a century.

The company had a reputation for strong day-to-day management. The company spent over $200 million, 5 percent of sales, on research and development. That percentage was, in fact, deceptively low. Bristol had a large consumer products business that required relatively little R&D. In actuality, 75 percent of R&D poured into pharmaceuticals; that amounted to about 14 percent of drug sales. Shareholders bene-fited from Bristol's prosperity. Each received a dividend of $1.50 for each share owned. And the shares—250 million common shares and 10 million in preferred, or nonvoting, shares—increased in value as steadily as an annuity. Every common share was backed by $3.45 in earnings, a forty-five-cent increase over 1983, and $15.55 worth of assets, another record. The stock hung around fifty for much of 1984, making its total value about $12.5 billion. What made Bristol's 1984 results all the more impressive was that it managed all this despite a newly robust American dollar that ate away at considerable overseas sales.[1]

Bristol-Myers had large, far-flung operations. It had 35,000 employees squirreled away in offices, factories, and laboratories all over the globe. It had 42,569 shareholders. To most observers, Bristol-Myers resembled one big drugstore shelf. The company sold analgesics such as Bufferin, Excedrin, Comtrex, and Datril; antiperspirants such as Ban and Mum; hair lotion such as Vitalis; Clairol hair-coloring products such as Nice 'n Easy, Loving Care, Born Blonde, Naturally Blonde, and Frost & Tip; and a variety of other products ranging from No-Doz tablets to Windex, Drano, Endust, and Behold to Son of a Gun hairdryers, O-Cedar mops, and True-to-Light makeup mirrors. It ran on like an inventory of the American Dream or a broom closet of mythic proportions. For the most part, these were products that marched off supermarket shelves briskly, and Bristol would never go broke selling them.

But while these generated a steady flood of cash, they did not have the capacity for the kind of rapid growth that could drive earnings, increase Bristol's profitability, and enhance the stock price. Many of these products were mature and many of the major markets—the United States, Japan, and Europe—saturated. Product extensions, say a new lemon Windex or a new hair color, would not help much. More-over, in the crowded, promotion-heavy world it was notoriously difficult to come up with new branded products. Bristol needed a whole new *business*, one that promised growth, one that produced, hopefully, large annual profits.

The Best Big Business in the World

Bristol had built such a franchise in cancer. The man behind it was Bristol's distant, patrician-looking chairman, Richard Gelb. Ironically, Gelb and his brother Bruce, who recently retired as vice-chairman, arrived at Bristol after their family company, Clairol, was acquired in 1959. Clairol was a triumph of marketing, not technology. By 1976, Gelb had risen to the top and begun to recast Bristol. For a decade he had been investing heavily to build an anticancer business. And indeed, Bristol-Myers in that time had become the Abbott Laboratories of anticancer, or chemotherapeutic, drugs. In 1984, Bristol sold five of the top ten anticancer drugs, generating $150 million in sales and a 40 percent share of the market; that was, like Abbott in diagnostics, over twice the share of its nearest competitor. Compared to a major blockbuster, say SmithKline Beckman's antiulcer drug Tagamet, which was approaching a billion dollars in sales, the anticancer market was nothing special. But it was a growth field: the market had been growing at a rate of 25 percent annually since the early 1970s. And very liberal pricing—in 1983 anticancer drugs increased in price 24 percent, over twice the rate of the rest of the drug business—helped. More importantly, anticancer drugs were becoming more precise and better understood. In 1984 analysts predicted that anticancer products would contribute as much as $600 million to Bristol sales by 1989, far outstripping Ban or Endust or Windex. For all of that, odds were if you had heard of Platinol, Mutamycin, Vepesid, Blenoxane, or Cytoxan—that is, Bristol's big anticancer drugs—you either were an oncologist or you had recently had a brush with one.

Anticancer drugs were not a glamour field. They never boasted the prestige of, say, the interferons. Developing and selling so-called cytotoxic chemotherapies—*cytotoxic* meaning "cell killers"—was a business somebody had to get rich off of. It was almost a public service; and Bristol played on this by funding a program of cancer grants and conferences. Cytotoxic chemotherapies lack the selectivity that Ehrlich sought in an effective pharmaceutical. No one has yet discovered an essential cellular function, or even the appearance of consistent antigens, that would significantly distinguish tumor cells from normal cells. Tumor cells do proliferate faster than most normal cells, but it is a difference in degree, not kind, making the attempt to exploit it highly problematical.

Anticancer drugs are poisons. They kill cells, usually by blocking some necessary cell function. They chop up DNA, say, or interfere with protein synthesis. Like a wrench jammed in a bicycle wheel, they stop cell division. The hope, of course, is that because cancer cells multiply faster than normal cells, a physician can kill them off before the patient succumbs. And that is why patients on chemotherapy grow deathly ill, shed weight like winter coats, lose hair, and flounder in an endless, nauseous sea. They are, in a very tangible sense, dying from both the disease and the cure.

Chemotherapy is also not a magic bullet; it is more a sawed-off shotgun blasting down a dark alley. There is nothing magical about radiation or about surgery, the oldest, crudest, but still most prevalent form of cancer therapy. But radiation carries an aura of humming high technology, and surgery has the surgeons, steely eyed, articulate, and rich. "Cytotoxic chemotherapy has always been an intellectual stepchild," recalls Dr. Stephen Carter, an alumnus of NCI and head of cancer drug development at Bristol-Myers.

> In the sixties, when I was in training, it was considered something only a few people did. Those terrible poisons. Those terrible, terrible poisons. [He shakes his head.] Chemotherapists were on the defensive. There was no medical chemotherapy as a subspecialty. It was a mixed bag of people that came together. When I joined NCI, I worked within a group of scientists who carried Burkitt's lymphoma and choriocarcinoma and childhood leukemias [cancers with which chemotherapies had considerable success] as a kind of talisman, as if to say, "See, it *does* work." Chemotherapy always suffered the criticism that it was not scientific, that most of the drugs were discovered serendipitously, from empirical mass screening for cytotoxicity. We were very, very defensive. With the successes of the late sixties, medical oncology exploded as a specialty. But still, chemotherapy has never come out of the criticism that it was an extremely unscientific, toxic approach to treating cancer.[2]

Cytotoxic chemotherapies, like nuclear energy, were hard to love, but difficult to do without. There was nothing that came close to being as effective against certain tumors. Not that Bristol did not sell other pharmaceuticals as well. The company was heavily involved with antibiotics such as Ultracef, Precef, Amikin, or Cefadyl; antidepressants such as Desyrel; and a cholesterol-lowering agent called Questran. Many of these were older products. In 1984, Bristol was eagerly awaiting approval of an antianxiety drug called BuSpar, which looked like Bristol's first major blockbuster. Unlike Hoffmann-LaRoche's Valium and Librium, BuSpar relieved anxiety for the so-called nearly neurotic without addictive side effects, interactions with alcohol, or Valium's famous buzz.

Bristol-Myers had submitted BuSpar to the FDA in 1982. The market enthusiastically ran up the stock, factoring into the price the $500 million or so BuSpar was expected to produce for Bristol following approval. The market, in its optimism, automatically figured this would happen soon. So everyone waited. And waited. And waited. Still the FDA did not act. Bristol went back to selling Endust and O-Cedar mops and figuring out ways to convince more oncologists to use more Platinol, which was going off patent (it finally lowered the price). It continued to build a massive billion-dollar research facility in Connecticut. It continued its cancer grants. Perhaps the waiting sharpened its sense of peril; certainly, it was irritating. But Bristol-Myers, unlike Genetic Systems, could shrug off the often-lengthy ponderings of the FDA without suffering unduly. Such were the differences between a big company and a small one.

Bristol-Myers had as much as anyone to lose to biotechnology. While Abbott was building a new business, Bristol was defending hard-won turf. Still, Bristol struggled to gauge some sense of the threat that biotechnology, particularly the immunotherapies, posed. Publically, spokespersons like Carter continued to argue that cytotoxic chemotherapies, like Platinol, would dominate anticancer therapeutics for the foreseeable future. Was the company simply slow to act, or was its analysis correct? Was it, in effect, defending its traditional markets, or was it really serious? Bristol was criticized by some analysts for not moving to license an interferon in the early 1980s or acquire antibody capability in the mid-1980s. And the company remained steadfastly unmoved when oncogenes—the latest "breakthrough" to threaten the cancer status quo—became a cause célèbre in 1984.

Oncogenes, far more than interferon or antibodies, were a classic product of the academy. Oncogenes were not for everyone; they were an acquired taste; they required, like certain abstruse literary theories, a certain amount of sophistication to savor. They never produced the grassroots excitement of interferon and antibodies, perhaps because they were more abstract, theoretical. And yet similar exaggerations and distortions clung to them. Oncogenes, in fact, were less a cure than an explanatory mechanism for cancer, although, of course, a mechanism might suggest a cure. Even simplified, oncogenes were complicated; oncogene investigators had about themselves the spirit of explorers on the far, frigid frontiers of biology. The trouble was, by the time the oncogene thesis began to reach the outside world in any big way, say 1983, it was becoming as intricate as something dreamed up by Rube Goldberg; it was receding further from the possibility of easy application.

For the traditional biomedical establishment, oncogenes were a tonic. They allowed the universities and cancer research centers to compete for attention with the biotechnology companies; it was proof that groundbreaking work in molecular biology was still being done. This was particularly important to fund raisers, administrators, and laboratory chiefs fighting the endless battle for funding. In the years since Genentech, they had felt increasingly beleaguered. Biotechnology had stripped many academic departments and cancer centers of their most luminous talents. The departure of Nowinski and his team from Fred Hutchinson, for example, was the first of a series of defections from that center; a small biotech industry subsequently grew up in Seattle, leaving Hutchinson, at least for a time, stripped of talent. And, of course, the best and the brightest were leaving NCI. Moreover, biotechnology was widely viewed as skimming the cream. Universities would train the talent, only to see it spirited away. (The fact that this argument was even raised pointed toward the ascendance of research, over education, as an income-generating enterprise at many universities; it was rather like universities bemoaning that pro football was somehow exploiting them by signing their still eligible players.) Finally, there was a widespread fear that the money coming out of NIH to the regional institutions would continue to decline.

As a result, the commercial constraints, crumbling before, truly began to be swept away. These ranged from the straightforward policy decisions—the more careful control and commercial exploitation of patents—to more controversial large-scale wooing of major corporations. The shift in perceptions was triggered both by a fear of Reagan Administration austerities and by the Reagan doctrine of privatization. "First, scientists and university administrators feared (unnecessarily, as it turned out) that Reagan would drastically cut budgets for all scientific research," writes Martin Kenney. "Second, Reagan changed the climate regarding the acceptability of industrial participation in public activities. The new ideology was that any activity that could be privatized should be."[3]

Against this backdrop unfolded the oncogene. Todaro and Heubner had hypothesized oncogenes at NIH in 1969. The discovery of a resemblance between certain viral oncogenes—genes in viruses that cause cancer in animals—to certain human genes took place in 1978. A year

The Best Big Business in the World

later, Robert Weinberg at MIT found that those genes could cause mouse cells in culture to act like tumor cells.[4] Activated oncogenes— genes that had switched on, like a short-circuited door bell—began to look like that elusive, essential difference between cancer cells and normal cells. Optimism soared: At the very least, if certain genes acted as cancer switches, could they not be used as a diagnostic? Or why could one not develop drugs to block their products?

Oncogenes satisfied the urgings of the powerful and prestigious virology crowd, many of whom had cut their teeth at NCI in the 1960s. It provided a molecular explanation, a mechanism for cancer. It was linked, chain by chain, to the most elegant and brilliant breakthroughs in modern biology. Given the world view of molecular biology, it made sense. New oncogenes were found, and their protein products were unraveled. Some of these genes coded for protein growth factors, like the kind Todaro focused on at NCI; others seemed to code for receptors for these growth factors on the cell membrane; still others seemed to code for so-called tyrosine kinases, important metabolic reactions within the cell. The bits of evidence seemed to fit together: a pathway, a circuit, in which activated oncogenes trigger increased production of receptors, growth factors, or tyrosine kinases, resulting in the cell spinning out of control. Thus results autocrine growth—the molecular basis of cancer, the unity beneath the diversity. At first, oncogenes seemed to work like a simple circuit made in a seventh-grade science class: wire, switch, dry cell, light bulb. Each cancer cell had one switch, perhaps even a distinctive one depending on the cancer type. But soon more and more oncogenes popped up. *Myc, ras, erb, sis, abl*, some subdivided alphabetically like an outline to *Summa Theologica*. Oncogenes, like the varieties of interferon, began to seem like distant cousins to the crowded taxonomy of particle physics. As time went on, the theory kept adding switches. Did they all have to be depressed—or just a few? Did they have to be depressed in any particular order? How many more existed back there in the genomic gloom? Was there a correlation between tumor type and the pattern of depressed switches, that is, the activated oncogenes? And how exactly did it happen? How could it be reversed?

By 1985 twenty-odd oncogenes had been discovered, with more to come. A few genes even appeared that could not strictly be called oncogenes at all but seemed important in the cancer process; that is, they lacked viral mates. Then appeared genes that might cause cancer by *not* turning on. Some researchers wondered aloud: Are the activated oncogenes a cause of cancer, or a symptom? After all, cancer cells are hotbeds of genetic mutation, of heterogeneity. That very propensity for change is one reason why cancer cells are such nasty customers. Cancer cells of certain tumors, for instance, can develop a frightening resistance to chemotherapy. Often, the drug will effectively shrink a tumor to a mere shadow of its former self. Then, months later, the disease suddenly returns with swift and murderous vengeance. Why? Well, the oncologist would admit, the drug did kill almost all of the cancer cells. *Almost*, alas, isn't good enough: the surviving cells, having developed the means to resist the drug, create a far more formidable adversary. Cancer cells, in this sense, operate at a far greater rpm than normal cells, like a speeded-up record. With their rapid proliferation, and the continual reshuffling of genes, they throw up a tremendous number of variations over a short period of time.

Thus, that evil word *heterogeneity* appeared. If cancer cells are such a mass of mutating, heterogenous biology, why are all the experiments performed on homogenous cell lines, born and bred under the fluorescent lights of the lab? The answer is, it's easier. You simply cannot experiment on people. And even if you could, the body is too unfathomably complex to know where to begin. So you simplify. You develop a cell line that is fixed and well understood. And then you play with the cells in dishes, altering this, jiggling that, inserting some viral DNA and then blasting it with x-rays. You disturb the system and see what happens. And you try to figure out why. Todaro's 3T3 line has proven particularly popular in this game. Although it looks like a normal, everyday mouse fibroblast cell, it also grows indefinitely like a cancer cell.

Thus the charge: 3T3 cells are models, artifacts, not the real things. Criticism of 3T3 led to questions about any cell culture formed from a single cell. Cells in culture do not swim in the complex wetlands of the body, but in an artificial environment bathed in an assortment of growth factors and nutrients—designer cells, in which you tend to get what you look for. Scientists could coax them to do certain things, like pump out viruses, but these cells did not really mimic the behavior of either normal cells or a heterogenous mass of tumor cells. They are neither natural or synthetic; they are a bit of both. Weinberg's great breakthrough, for instance, involved inserting an oncogene into 3T3 cells and watching them transform into full-fledged cancerous cells. How could he tell they were cancerous? Because they formed little piles, like bricks dumped on the carefully fitted tile floor that 3T3 cells normally construct. But could he do it in normal mouse cells? Eventually, Weinberg's lab succeeded at just that, using two oncogenes to trigger cancer in normal mouse cells. Still, said the critics, that did not mean that human cells would act the same way.

There were other issues. Was cancer triggered by the waywardness of a single cell or by the breakdown of a community of cells making up tissues and systems? Oncogene theorists drew the image of a single cell in which one, or a series of genes—normal, functional oncogene precursors called proto-oncogenes—are activated by environmental insult, like smoking or radiation, or by the worm trail of an invading virus. They begin to produce, to overproduce; they alter the developmental balance of the cell. One cancer cell creates two, which create . . . many. The Darwinian imperative takes over. The growing tumor squeezes out healthy tissue. The grim decline begins. For all of that, no one had ever actually seen the original, single shift from normal to cancerous except in experimental *in vitro* systems. One could as easily argue that tumors arose, like urban crime, from wider, systemic changes, far more involved than just a few normal genes shifting to new, dangerous chromosomal positions, or suddenly making a hundred copies of themselves like a narcissist at a Xerox machine. The sheer complexity suggested a more complex response than the oncogene thesis provided.

The point of all this? Oncogenes, even if they do one day provide the elusive unity beneath the heterogenous surface of cancer, were certainly not a commercial solution for much of anything—not yet, anyway. There were too many questions, too many criticisms, too many dark corners. In 1984, at a cancer meeting, one could still hear: "In five years, we'll have it all wrapped up." A year later, the optimism had gone flat. Even if oncogenes did prove to be the key, what could one do with it? Perhaps one could develop drugs that inhibited the growth proteins produced by oncogenes, although those proteins might have normal functions as well. It was not as if one could go from cell to cell, flicking off switches like a thrifty monk in a sprawling monastery.

By then deal makers had already capitalized on the oncogene boom— or, rather, boomlet. On Wall Street oncogenes appeared in company names as a simple metaphor, the cancer switch, and a prefix, *onco:* Oncogen, Oncogene Science, Oncor. A handful of limited partnerships and joint ventures were assembled, constructed around oncogenes. Significantly enough, once Oncogen was formed, Todaro himself chose to sidestep the high scholasticism of the discipline and opt for a bit more empiricism. He was not hunting for more cancer genes. He passed, as well, on developing diagnostics to various *ras* or *myc* genes—DNA probes—a big effort at Oncogene Science. Todaro viewed those efforts as too murky and the market too small. Instead, he was trying to use a number of the new techniques to manipulate the circuitry of cancer that had been revealed by the oncogene work. First, he focused on monoclonal antibodies against five major tumors: lung, breast, colon, prostate, and leukemia; plus the so-called transforming growth factor (or TGF) antigen, a candidate for the long-sought-after general tumor marker, the single detectable sign that a cell had turned cancerous. Oncogen could draw off the expertise accumulated by Genetic Systems. Second, he targeted the oncostatins, proteins that seemed to inhibit certain tumors. This line of work came directly from Todaro's work on an autocrine mechanism.[5]

Oncogen was thus a rare beast in biotechnology: combining monoclonal antibodies *and* recombinant DNA. "We thought we were somewhat unique," he said. "We were focusing on a disease—cancer—rather than a technology. Some of the cloning companies were really pure technology companies waiting to exploit a discovery made by someone else. We thought there was room to make the primary discovery, and hook up with someone else to do the marketing and manufacturing." On the other hand, Todaro looked on diagnostics much as Nowinski did: as way stations on the road to therapeutics. Diagnostics might be commercially important, but they did not make the hair on the back of the neck of the *real* researcher stand up. "The original idea [for Oncogen] was cancer diagnostics with Syva," recalled Todaro. "Then it became clear that what we were working on had as much to do with therapy as diagnostics. That's where my real interests lie. And that's what most of the people I recruited were interested in. You can't get top quality people and say, 'I want you to make a monoclonal but you can't think of using it for treatment.' "[6]

Todaro radiated optimism. He believed that the gap between science and technology, research and development, had narrowed and that truly effective products might come from proprietary, patentable molecules—monoclonals, oncostatins. The Cold Spring epiphany lingered.

At NCI he had gained a reputation as a scientist who needed to be near the action—sometimes, perhaps, to excess. Genial on the outside, he was intensely ambitious within. Unlike Nowinski, Todaro did not have to cater to Wall Street, although he did accompany Nowinski and Glavin on the investment circuit. He did not have to don the robes of the entrepreneur, although he occasionally tried them on. His ambition, preoccupied with scientific glory, did not seem to have room for irrelevancies like marketing and manufacturing. He was bringing NCI to

The Best Big Business in the World

Seattle: commercial, yes; optimistic, certainly. But even Todaro seemed to know he was a scientist, not a businessman.

And yet Oncogen cost a lot of money. Even at NCI, hardly a group obsessed by thrift, Todaro had a reputation as an enthusiastic spender. That followed him to Seattle. He had opened his checkbook to lure a Swedish husband and wife team at Fred Hutchinson, Ingegard and Karl-Erik Hellström: he was getting the best and the brightest minds in biology, and they did not come cheap.[7] He often simply did not seem to care about managing the enterprise, however. Ken Gindroz, who ran administrative affairs at Genetic Systems, recalled approaching Todaro about patents: Todaro had been using an outside firm at $7,000 a pop—and he stubbornly refused to switch to a less expensive inside patent counsel.

Oncogen was originally designed to run over four years on $9.5 million. But on the November day the new labs were opened, the two companies announced that they were raising the stakes. The project, with fifty employees (the figure would rise to sixty-five by year end, with twenty-four Ph.D.s) would now be getting $14 million. There were, at the time, two ways to take such an announcement. Optimistically, one would suppose that the companies, particularly Syntex, the major backer, had grown excited by Oncogen's prospects. Pessimists would say that the project was going to cost a lot more than anyone expected.

Enthusiasm ran high. Todaro had gotten off quickly. This was the kind of project Nowinski loved. And a few million dollars made Syntex's efforts to get access to some remarkable researchers worth it. On the other hand, as Glavin said, "The further we got into it the more expensive it seemed."[8] Syntex now agreed to pay $10 million instead of $8 million; Genetic Systems, $4 million instead of $1.5 million. For Syntex, a couple of million dollars a year was no big deal. But Genetic Systems had no appreciable cash flow, and the million dollars a year it had to pay to Oncogen—up from $375,000—would further depress, if not devastate, its bottom line. In the short term, it would make losses a bit deeper. In the longer term, it would delay the day of profitability, unless Todaro provided that major product (although even then, Syntex would get the bulk of the early profits).

Meanwhile, Bristol-Myers waited. If one listened carefully, the company was not saying that the biotechnologies would never have a role to play, just that it would take time. If that perspective was taken seriously—and not just as some lame defense of its own hardheadedness—Oncogen had much to offer. Todaro had a brilliant record; he had assembled a fine research staff; he had built a research organization focused, like Bristol's own laboratories, on a single disease, not on a single technology. He was using the insight brought forth by the on-cogene, but with a slightly more empirical air. Certainly, the potential products were untried, untested molecules that faced a long and rig-orous period of basic research, of testing and FDA consideration, and of patent squabbles—the same old story. But Bristol, unlike Nowinski, was in no hurry. The fact that Todaro had built a little academy in commercial garb in Seattle, a little bit of NCI along the sound, looked like an opportunity. In the past, those sorts of organizations were not available for purchase. Now they might be.

CHAPTER 16

The Rocky Commercial Road

THE GENETIC SYSTEMS 1984 annual report was a glossy, handsome affair, overseen personally by Isaac Blech. It was full of photographs of glowing fluorophores, clean rooms, and cancer cells. A group photograph showed department heads gathering next to a ficus tree rising through the atrium. Nowinski grinned widely, like a proud father. A few pages later, George Todaro smiled as well, his photo inset against an ominous blue microphotograph of a cancer cell. The report conveyed the conventional optimism: "All of Genetic Systems' resources—intellectual, technological and financial—are working together in a strong, steady pattern of growth."

There were hints of discord for those who looked more closely. In the 1983 report, a year earlier, a photograph had shown Nowinski, in sweater and loosened tie, explaining plans for new manufacturing facilities to Glavin, whose face and tie were both as tight as a fist. A year later, while Glavin's name still sat atop the organizational chart as chairman of the board, no photograph of or reference to him appeared.[1] He had, for all intents and purposes, ceased to exist. Why this disappearance? The financial report, in the back of the annual, told part of the story: Glavin's role declined as the financial picture darkened. Losses had widened far beyond what analysts, not to say Nowinski, had foreseen—although, of course, the report did not mention that fact. While revenues grew to almost $9 million, nearly all in the form of contract

155

revenues and interest on bank deposits, the company posted a loss of almost $4 million, $2 million more than the previous year. The company's per share losses had gone from eight cents in 1983 to *twenty* cents. Its bank balance had shrunk from almost $37 million dollars to just over $30 million.

By then, early 1985, the knock in the Genetic Systems engine had developed a rattle and a hint of smoke. The AIDS test and the Applied Biosystems joint venture indicated that Nowinski had to bet more to save the game. Product sales were not materializing. Oncogen, which would not have products for some time, was draining cash. The passage of time had become ominous. All these factors went far beyond Syva's difficulties with Abbott, or with proliferating competition in diagnostic markets. Rather, it had become more obvious that Genetic Systems' inability to turn a profit stemmed from a fundamental decision made very early on: the emphasis on infectious diseases. It was not only that Genetic Systems and Syva were not making much money in sexual and respiratory diseases—no one was, except Becton Dickinson, the biggest producer of traditional petri dishes and culturing media. Abbott was not; neither was DuPont nor Johnson & Johnson. For all the studies and all the forecasts, the microbiology market refused to accept new products in large quantities. It was baffling. Certainly there was a need. Certainly the technology made testing considerably faster, cheaper, and easier. Certainly these tests, unlike cancer monoclonals, actually worked. Didn't the customers—the lab technicians, the administrators, the purchasing agents—know there was a revolution going on?

What went wrong? A fatal combination of technical inadequacies and marketing misjudgments. The most obvious problem was also one of the most intractable, and it affected both the traditional method of growing bacteria in culture dishes, then testing them against antibiotics, and the new immunoassays. Say a woman walks into her gynecologist's office feeling ill. The doctor suspects either gonorrhea or chlamydia. But there are few physical signs of it, unlike with the male who wears his affliction more visibly. The usual method of collecting bacteria involves three samples: from the vaginal canal, the cervix, and the anus, using a swab left at each site for a minute or so. But in this imperfect world, physicians, eager to get the job done, tend to skip sites or hurry through the swabbing. In the case of chlamydia, the bacteria often hides in the mucosa, high up in the vaginal canal. To get a usable sample, the mucosa has to be scraped, not just swabbed. Even when correctly done, the sample often rubs off as the swab is pulled back through the canal.

The failure to get a good sample made marketing new, faster, but more expensive tests more difficult than the market research studies indicated. Marketing involved more than just the ability to sell, like a Fuller Brush salesperson on the front stoop. With products like the infectious disease immunoassays, it required the ability to understand, master, and use the dynamics of a marketplace to sell a product.[2] Marketing, in this sense, consistently took a back seat to science at Genetic Systems, just as Glavin steadily lost ground to Nowinski; his ability to reject what he had earlier referred to as "superficial and facile generalizations about marketing a product" was eroding. Nowinski, it is true, knew microbiology. But he knew it as a researcher, as an academic, not as a civil service technician in a public-health facility slogging through endless tests.

This was unfortunate. Diagnostic markets were complex and volatile, and the concept of the "customer" was an elusive one. It was very different from pharmaceutical marketing. A sales representative from a drug company normally sold his or her wares directly to the physician. This was not easy, but it was straightforward, because the physician actually wrote the prescription. For the diagnostic salesperson, the "customer" was a fragmented concept. The actual buyer might be a laboratory administrator, who, in turn, rubber-stamped recommendations made by the heads of eight or nine different laboratory departments: clinical chemistry, microbiology, serology, hematology, coagulation, stat labs, special chemistries, microscopy. The department head, in turn, sought to cater to her nominal customer, the physician, who ordered tests, after *his* customer, the patient, trudged into his office suffering from one ailment or the other. Like a small army crossing a long border, a diagnostics company had to chose which ground to attack. Abbott, for instance, made its move in one of the larger departments, the clinical chemistry lab. But even that market took years to open up to new testing methodologies. Microbiology, which Genetic Systems specialized in, proved to be even more stubborn.

Part of that problem could be traced to the bureaucratic rivalries within the diagnostic laboratory. For instance, therapeutic drug monitoring was the classic test for the clinical chemist. The purpose of drug monitoring was to compare the level of common drugs found in patients' blood against standard curves; it was the kind of test that might, given the proper technology, be performed on every patient receiving drugs, every day. If the level was too high, the physician would reduce the dose to the patient; too low, he would increase it. The clinical chemist thus required quantification: How *much* is too high?

Microbiology, however, was different. Instead of identifying the amount of a drug, the microbiologist labored to identify bacterial or viral presence. Thus arose the practice of growing up microbes in culture until they formed identifiable colonies. Physicians were less concerned with which microbe had infected their patients than with a more practical question: Which drug kills it? And so microbiologists would culture the bacteria, plant them in dishes larded with antibiotics, then check back later to see which antibiotic was most effective. In this sense, traditional microbiology culturing was as empirical as drug screening or chemotherapies: not very scientific, but . . . *it worked.*

The immunoassay altered this state of affairs. In the 1960s polyclonal antibodies were linked to radioisotopes as markers for the first time. Unfortunately, this meant that they could only be used in radiology departments with Atomic Energy Commission clearance. The technology was exquisitely sensitive, if cumbersome, but the radiologists had not had much experience, and did not show all that much interest, in *in vitro* testing. But which lab section would inherit the tests? The natural candidate was clinical chemistry. Radioimmunoassays were so sensitive that they could quantify exactly the amount of a compound in a vial of blood. Besides, many of the chemistry technicians had had some training with radioisotopes. Thus, by the late 1970s, immunoassays were flourishing in the clinical chemistry departments, providing an opening for Abbott.

But immunoassays continued to evolve. Safe, easy-to-use, nonradioactive enzyme and fluorescent tags soon replaced radioisotopes, and the issue of who would get the tests arose again. Take the case of human chorionic gonadotropin, or HCG, a protein that in women indicates pregnancy, and in men, at certain levels, the possibility of testicular cancer. When pregnancy was suspected, serology would test for the *presence* of HCG in blood serum. This was not a quantitative test at all. Cancer testing, on the other hand, required exact levels to be calculated, making it the province of the chemists. Microbiology was caught in a similar conflict with chemistry when it came to infectious diseases. Like serologists, microbiologists traditionally searched for presence. But chemistry began to infiltrate their turf with its ability to quantify. They could now provide the answer to the question: How *many* flu bugs are swimming in that bit of blood? Meanwhile, the doctors were in flux, too. Most were still satisfied with knowing what bug was in the sample, in order to prescribe the correct antibiotic. But as antibiotics became more specific, some physicians, worried about bacterial resistance to antibiotics, began to call for more quantitative tests.

All this created a dilemma for the testing-laboratory microbiologists. The most sophisticated tests they had were the Microtrak tests from Genetic Systems and Syva, but even those tests, while innovative and interesting, did not really solve their problem. Although Microtrak could quickly identify presence—all those fluorophores glowing under the microscope—it could not indicate antibiotic effectiveness. So researchers had to culture anyway in order to get enough bacteria to identify type through a traditional microscope. Besides, as budgets tightened, there was the issue of buying a bulky, expensive fluorescent microscope required by Microtrak, particularly when no one knew how much new business would be generated. So why not just forget the whole expensive affair and stick to culturing?

It was a case of good science packaged inadequately—of good technology that was not good enough. An effective micro test really had to do three things: be sensitive to bacterial types, indicate drug effectiveness, and minimize collection problems. None of the products on the market in the mid-1980s could do all three. As a result, microbiology tests languished, although Abbott began to make some headway by bypassing microbiology altogether and selling easy-to-use strep and gonorrhea tests to physicians, and Becton, king of culturing, began to sell polyclonal systems to its old laboratory customers while continuing to profit from the culture dish and agar business. Still, said one diagnostic marketer, "We don't have a product yet that is truly marketing viable. I think it's almost an unjustified criticism to say this market hasn't taken off yet, because a product really hasn't appeared yet. Who's going to run a bastard, bitchy test with a high false-negative rate? I think it could take another ten years to really grow it up."

The refusal of the infectious disease market to open created a variety of financial repercussions at Genetic Systems. In March, the Class B warrants had been extended a full year, to June 1985; the share price had fallen so low that shareholders weren't exercising their option to buy new stock. In May the company announced plans to raise from $25 million to $35 million in a second limited partnership. This was not an unusually large amount. In 1983, with Wall Street hot for biotech in any form, Cetus had sold a partnership funding a combination of therapeutic and diagnostic products; it raised $75 million. Genentech had sold its second partnership that year for $32 million to fund a heart-clot dissolver, tissue plasminogen activator. Even with the market crumbling in 1984, Biogen managed to raise $60 million, and, significantly, Hybritech $70 million.

A year later, investor confidence was shaky. Low stock prices scared off the kind of private investors who would buy into limited partnerships. The tax laws were also in flux; Congress talked of taking away the tax advantages of the partnerships. In October 1984 the company announced that its was scaling back the partnership to $22.5 million. Bundled in the package now were the AIDS test, the Applied Bio analyzer, now dubbed Chemware, and some thirty-five monoclonals against cancer. To the investors, formally called Genetic Systems Diagnostics Partners, Genetic Systems offered tax breaks, royalties of up to 12 percent if both reagents and instruments were sold, warrants in Genetic Systems and Applied Bio, and up to $4 million back in the event of cost overruns. Moreover, partners would also share in a joint venture to manufacture the product. All in all, this was a generous deal for investors. For Genetic Systems, on the other hand, profits had to be shared with an extraordinary lineup of partners: with the limited partners; with Applied Bio on the Chemware; with Diagnostics Pasteur and its marketer, American Hospital Supply, on AIDS; and with Syntex on the cancer tests. Such a system was like carrying water in a leaky jar—one had to move very quickly or possess a very large jar.

Nonetheless, Nowinski was enthusiastic. He argued that the AIDS test, run on Chemware, represented a leap from research house to integrated diagnostic company. This, however, was the palest flicker of true integration. American Hospital would do the marketing; Applied Bio would share in the hardware; limited partners would get half the manufacturing revenues. Although the deal *would* provide immediate cash and the opportunity to take another swing at a major product breakthrough, it would require huge sales to generate even modest profits, after everyone else had taken a cut. Nonetheless, even before the partnership closed, Genetic Systems was pumping money into the AIDS program. By the end of 1984, 205 units out of the projected 340 had been sold at $50,000 each. The company, in turn, received over a million dollars. Not all of this would remain in house. Genetic Systems had to pay Diagnostics Pasteur $250,000 for the rights to the technology; and it agreed to pay another $750,000 over the next three years. That cash would, as well, come from later payments made by the limited partners.

On December 19, 1984, Bob Nowinski wrote to shareholders: "I am pleased to report to you several significant recent developments at Genetic Systems associated with the expected introduction next year of major products including tests for AIDS, hepatitis and Legionnaires' disease. In anticipation . . . the company has made moves to strengthen its management, financing and technology base."[3] More specifically, Glavin was kicked upstairs to be chairman of the board. Nowinski took on the position of chief executive officer as well as scientific director and announced the hiring of a new chief operating officer and president: Joseph Ashley.

Ashley was widely considered a diagnostics instrumentation whiz. He certainly knew how to package diagnostics technology into products. He had played a major role in building Beckman Instruments into the most successful diagnostics company going, back before the rise of Abbott. Ashley had helped shepherd sales from $11 million in 1974 to $400 million nine years later. That was the year the company was acquired by Smith, Kline & French, a Philadelphia drug company suffering from a serious drug-research drought. In 1984 the hospital market fell apart and Beckman staggered. SmithKline (S K & F renamed itself SmithKline Beckman after the acquisition) reacted by cutting budgets. Without new products, Beckman lost all momentum and crashed. Ashley, disgusted, settled into retirement—at least until the investment bankers at Morgan Stanley introduced him to Nowinski, who painted an image of exciting, entrepreneurial Genetic Systems. The company, said Nowinski, was ready to break into the marketplace. Ashley thought he had the key to making it in diagnostics: pumping out product. Nothing else mattered as much as the ability to generate product.

Again, Nowinski had shown a certain touch. There were few executives like Ashley available. Tall and thin, Ashley seemed to run on nervous energy; he was also confident and very tough. One analyst who visited Seattle scribbled in his notes: "Ashley: Smooth, competent, independently wealthy . . . but does he have the stomach for the long haul?"[4] An electrical engineer, not a scientist, Ashley knew how to build hardware and he knew what the market required. He had a realistic view of the market. "He did a lot of development work himself," said Howard Teeter, who had been Ashley's boss at Beckman. "Ashley had an insight into instruments that was extraordinary and a feel for whether an instrument is useable, will sell, will fit a need. His ability to psych that out was awfully good." Ashley, said Teeter admiringly, "was a tightwad manager. He could sometimes be hard to get along with. He was the kind of guy you'd get if you wanted to make money. If you wanted to run a Social Security company, or you wanted to do things just for the fun of doing them, he may not be the guy and I might not be the guy. We never worked that way at Beckman."[5]

Despite the entrance of Ashley, analysts began to lose their patience. It was difficult to argue against the market. Many of the analysts from the early days had drifted off as the market sank. Nelson Schneider, for instance, had become a venture capitalist. Others were laid off as Wall Street firms cut back. A new group slowly took their place, and they started off fresh. They were not anxious to back what they perceived as losers. "Genetic Systems was one of the few companies I did not do an earnings model on," said one analyst. "How can I actually think anyone is going to make money investing in Genetic Systems with its current valuation? With 23 million shares outstanding, to earn a buck a share on the bottom line would imply that they do over 200 million in sales and have a 10 percent margin after-tax. That's almost as large as Abbott Diagnostics. So you ask yourself: Is this another Abbott? And you walk away, shaking your head, and say: No way."

Nonetheless, Nowinski was far from finished. If money and credibility were required, he would find them. Deal making could paper over any problems that cropped up. Annual reports for companies on a calendar year usually appear in early March or so. This one was delayed by a deal that, if nothing else, brightened the gloomy reality of the financials.

On February 7, 1985, Bristol-Myers announced that it was buying a third of the shares in Oncogen for almost $13 million. In all, Genetic Systems, Syntex, and Bristol were increasing their total investment in Oncogen to over $20 million over the next three years. Negotiations between the four organizations had been going on for some time. "Even Syntex realized we were getting into a lot of bucks, which is how Bristol-Myers came in," said Glavin.

Syntex was willing to take on another partner and Bowers [at Syntex] knew William Miller, president of Bristol. The negotiations went from top to bottom. The business aspects were not the thing that made all that much difference. The science mattered. All the program heads [at Oncogen] gave presentations. What they were doing. Timetables. Pretty forthright stuff about what you have got in hand and what's speculative. At one presentation we had six talks. Todaro was running it. The last guy in is fresh off the plane from Europe. Leather jacket. Jeez, I say, George . . . But when his turn came, he got up, without anything prepared, and went to the blackboard and started talking. He was the best one. He really impressed them. He spoke so clearly.[6]

It was a triumph for Todaro. He had doubled his budget in two years. He had the credibility of Bristol-Myers—and its mighty chemotherapy program—behind him. He had the freedom that having three bosses can create; he could play one off the other. Bristol did manage to install a business manager at Oncogen named Brad Simmons who had worked for Abramo Virgilio, Bristol's head of science and technology. Why not? Bristol was contributing the most money. Todaro did not alter his operational style after Simmons arrived.[7]

Indeed, it looked like the perfect deal. Syva craved diagnostics; the bigger the budget, the faster Todaro and his troops could, in theory, develop new cancer tests. Bristol-Myers did not have a diagnostics product line. Syntex was willing to concede cancer therapeutics—linking monoclonals to Platinol or Vepesid, for instance—to Bristol. Where did that leave Genetic Systems? A junior partner in a tough club—but a partner nonetheless. Oncogen would constitute a cash drain, but Genetic Systems would benefit from contact with Todaro's whiz kids. You never knew what serendipity would throw up. The down side, however, was downplayed: the fact that it further reduced the possibility that Genetic Systems would be getting any significant revenue from Oncogen—after Bristol-Myers and Syntex took their cut—any time soon.

CHAPTER 17

Permanent Revolution

AS LOSSES WIDENED, Genetic Systems stopped being such a fun place to work. Employees began to drift off, voluntarily or otherwise, and tensions increased. By 1985, Nowinski had gone through one operating man, Glavin, and one administrative head, Max Lyon. Managers like Lyon, in particular, had a rough time dealing with him. They lacked scientific cachet. They dealt with areas Nowinski cared little about: offices, supplies, support personnel, janitorial services, accounting, planning. Necessary stuff, but it did not take a Nobel Prize winner to get them done; and Nowinski was not good with people he did not respect.

Management was where bureaucracy, from the perspective of an entrepreneur like Bob Nowinski, began to gum up the works. Every company, whether an entrepreneurial one or a large corporation, has to wrestle with achieving a balancing point: between freedom and control, flexibility and structure, and the tangibility of the present and some fantastic, if still unformed, future. The tendency in large corporations is to drift toward structure and its human manifestation, bureaucratic management. For the entrepreneur, the tendency is to emphasize freedom, creativity—or creative chaos—at the expense of stability and control. Nowinski's managerial style reflected certain tendencies within any entrepreneurial company. Although many of the specific difficulties were his own, they also tended to be the kind that generally afflicted the biotechnological entrepreneur.

Bob Nowinski did not seem to accept the need for structure. The technology was so revolutionary that it required a new kind of organi-

164

zation to bring it to life, and only he knew what that might entail. He acted as if the bureaucracy did not exist—or, more precisely, that it existed at his whim. He refused to recognize limits on his power. He operated as a free agent, wandering into situations, snapping off decisions, then wandering off again. "He acted," said a former employee, "as if everyone worked for him and that was that. He dealt with everyone on a personal basis—and he could be extremely vindictive personally. There were good guys and bad guys. Nobody was immune to it." He also proved quick to judge, a darker side of his ability to learn quickly. This applied even in the lab, where he leapt to conclusions so rashly that some of his scientists, his natural constituency, began to wonder if he had lost touch scientifically. His credibility suffered. His talents—charisma, creativity, energy, the ability to master a subject quickly—were perfect for deal making, the restless, relentless construction of new relationships. They, and he, were not so good with the management of old ones.

He was, indeed, a deal maker at heart. He seemed to search continually for the new deal, the new relationship. The Bristol-Myers deal on Oncogen was the kind of complex, inventive deal he loved—an architectural blend of science and Wall Street, of money, power, and high intellect. Strategy and management, from this angle, are tasks fit for those noncreative, occasionally necessary drones, the managers. Managers grind away at their task, every day; they live with limits, they seek less transcendence, than goals. To the manager, force is what matters: How much force in the form of cash or bodies can be brought to bear upon a problem? To the deal maker, the world appears as if it can be conquered by the exercise of will, by the leap. He or she traffics in ideas—believing that ideas can be willed into reality—whereas the manager hauls around the solid stones of reality, hoping to rearrange them slightly. "Nowinski would have made a great investment banker," commented an analyst. "But a businessman? Not really."

In the best of worlds, deal makers balance managers, and entrepreneurs offset bureaucrats. Such an equilibrium exists only rarely and then, temporarily. And yet that was the surface Genetic Systems so expertly flashed to Wall Street. From the start, however, the reality lay elsewhere. Glavin struggled to manage the construction Nowinski had thrown together, but he did not come from a diagnostics background. He was immediately at a disadvantage. The organization grew wildly around his feet, but without a solid infrastructure, a controlling hand. Glavin finally lost control. Nowinski thrived where he could exert his will most easily. He gained power, paradoxically, as the business plan—

his plan, mind you—came apart at the seams. "Nowinski," said one former associate angrily, "knows everything. It doesn't matter what it is, he knows it. You start from there." The guys at Syva had it right: "No-lose-ski."

Nowinski used his inspirational gifts not only with investors but with employees. He sought a permanent revolution. All revolutions—entrepreneurial, social, or political—inevitably fade; revolutionaries grow stiff and cranky; people seek escape in small, conservative comforts; passion turns to dogma. Wiser heads in the drug industry knew this. By 1985, Genetic Systems was no longer a new company; veteran employees had been listening to Nowinski for five years now. It was too long. The treats, the retreats, the happy hours, and the staff meetings were becoming forced, like a party past its prime. Rather than relieve tension, encourage dissent, or create loyalty through participation, Nowinski increasingly leaned on personal force—inspiration, threats, or promises of wealth. His hand on the throttle, unlike Swanson's or Hybritech's Ted Greene's, was heavy. He was best with a new, impressionable audience for whom the rhetorical flourishes were fresh.

His charm did not last long with Ashley. When he arrived, he was immediately warned about Nowinski. "I thought to myself," said Ashley. "I've known tough SOBs before, here's another one. But I wasn't really worried. You work out a way to communicate and go forward. That's management."[1] Ashley set out to discover what lay beneath the surface of Genetic Systems. He was not surprised to find a predominately research operation; he had been hired, after all, to transform it into a business. He was more shocked to find that the sexually transmitted diseases tests—"the key to infectious diseases," he said—had been given away. "It worried me. At one point I called some outside bankers to tell them that I thought the store had been given away. And they said, 'Oh no, you're wrong. They didn't do that.' And I said, 'Please, I'm telling you. There's not much here.' "

More disturbing was the gap between Nowinski's view of the situation and that which came from below. Nowinski suffered from a growing credibility gap. Early in his tenure, Ashley sat down to hear a progress report on the Chemware system being developed with Applied Biosystems; Chemware was the key to marketing not only the AIDS test but many of the rest of the diagnostics as well. And Nowinski believed that the AIDS test was his ace in the hole. When Ashley arrived at the briefing, he was met by the senior scientist in charge of Chemware, Dr. Karen Hargraves. It was December 1984. Said Ashley:

Permanent Revolution

I was standing there and she came up to me. She said to me, "Do you really want me to tell you about Chemware?"

"Sure."

"The truth?"

"If you're going to get up there and give a talk about some piece of hardware, give me the truth."

She then started telling me about how scared she was [to tell the truth]. I told her Nowinski would want the truth and besides I was going to be there. Anyway, she got up and told the truth; the system wasn't working. He suddenly jumped up and tried to overcome what she said. In fact, he took over that part of the technical discussion himself. Now I suddenly got concerned because I recognized the play: Here are people doing the job saying it's not going to work, and the guy who needs it to work shouting how great it is. Then, after the meeting, he took her aside and beat the hell out of her verbally. I began to realize then that that was his technique. You don't step out of line.

Other problems surfaced. The gleaming 15,000-square-foot manufacturing facility that had cost $2.5 million to build and innumerable press releases to describe, was, in Ashley's view, junk. It resembled a pilot plant more than a factory. It had been designed by people who had no idea what pharmaceutical manufacturing really required. It had no space on the shipping dock for storing finished goods. It had no warehousing for inventory not yet approved by the FDA. It could crank out enough for, say, Legionnaires', but it would be useless if just one of Nowinski's forecasts came true. An outside consultant calculated that its capacity was just over three million tests a year; Nowinski was predicting that the company would soon be selling *sixteen* million tests.

Ashley went out and found an additional 30,000 square feet nearby. Almost immediately, he and Nowinski began arguing about how large AIDS would be and how much equipment should be bought and installed. Nowinski believed Genetic Systems could seize the AIDS market worldwide, thus requiring huge amounts of manufacturing capacity, anywhere from thirty to sixty million tests a year. He wanted all the equipment bought and installed, and the operators hired and trained and ready to roll the moment the orders flooded in. Ashley was less sanguine. He thought that Genetic Systems could never hope to beat Abbott over the short term, although, with a better test, it stood a chance, though slim, if it could get more financing; he wanted just enough equipment to meet immediate needs, say sixteen million tests.

167

Finally, the two compromised, and the plant was built with a capacity of twenty-five million tests.

The debate over manufacturing was only part of the escalating friction between Ashley and Nowinski. Ashley tried to handle Nowinski by taking the offensive. "It was very stressful," Ashley said. His strategy was to deflect aggression with aggression. Snap back when bullied. Correct him at board meetings. Paint a more realistic picture to restore credibility. Above all, get some products on the market. Ashley had certain advantages as a manager over Glavin: he had a reputation as a diagnostics expert—The Man Who Built Beckman—and as a professional manager; blunt, tough, and generally lacking in the awe for research that Glavin demonstrated and that Nowinski craved. He knew what he knew, and he was ready to challenge Nowinski. Ashley realized that Genetic Systems was in trouble. Disenchantment had spread. Everyone, save a few scientists who were interested in their work only, knew that the situation was deteriorating. Ashley found he had allies on the board. Glavin, before leaving, had succeeded in getting several outside directors—Dean Thornton, president of Boeing, and Donald Stenquist, president of a Seattle manufacturer called Criton Technologies—installed. Indeed, the board itself had urged Nowinski to hire an operations professional like Ashley; and this sudden interest in management created enough of a shift to block Nowinski and give Ashley some leeway in diagnostics. Those two outsiders, with Glavin and Ashley, could line up against Collinson from the Schroder, the two Blechs, and Blech ally Solomon Manber. The emotional swing man—although it never came down to a vote—was Dr. John Hansen, the medical director, an early colleague of Nowinski, but an independent force.

Ashley was, from long experience, obsessive about product flow: it was really all that mattered. One of Ashley's motivations behind the second manufacturing site was to move operational personnel there eventually, leaving Nowinski and his researchers back at the fancy building on the waterfront. Ashley was a manager; he installed procedures, organization, planning, and *structure*. Ashley brought in an accountant named Udo Henseler from Beckman as vice-president for finance—in effect, the chief financial officer—to make sense out of the accounting system. Henseler discovered that were no management systems at all in place—no purchasing system, no financial controls, no integrated computer system. If a researcher wanted to order a million mice, he called up a supplier. He might, or might not, tell someone in accounting. Eventually a bill might wend its way to accounting and get paid. Or it might not. It was a minor miracle that SEC disclosure documents were completed and filed on time. No one quite knew how much cash the company had on hand. And yet when Henseler began to install systems, reporting requirements, and information processing, he met immediate resistance. "It was culture shock," he said. "They had never had to do it before."[2]

"It had been unstructured and unorganized before Ashley," said Ken Gindroz, who arrived in June 1984 and replaced Lyon as chief of administration.

> Bob's style was unstructured. You would get halfway done with something and have to move off in another direction. We would do the planning for a new lab for a new project, and suddenly everything would change. The direction was always changing. Ashley, for instance, started having management meetings. When he first started, he would say, "Here are the products we're working on. Everyone agree?" Right, right. Then, next week, he would ask how everyone had done on their projects. And a scientist would say, "Well, I didn't work on that project, I started a new one instead."[3]

Nowinski did not, at first, sit in on those meetings. He resisted the attempt to impose structure by ignoring it. He would, for instance, send out press releases without telling anyone. When the calls would come in, nobody, including the investor-relations office, would know what to say. That, in turn, confused Wall Street investors and analysts. Projects would begin without anyone knowing they existed. Labs would be set up for one project, then have have to be rebuilt overnight for another.

But Nowinski was not alone in resisting the new regime. Nowinski's power base was still strongest among the scientists, particularly the ones tied to pure research. "Early on, you had a whole bunch of scientists who were used to academic freedom," said Gindroz. "They were used to changing direction on projects whenever they wanted, rather than the profit or results orientation you have from a business point of view. Intellectual curiosity—or what kind of neat thing I wanted to work on—passed for planning."

Ashley tried to move the organization from research to development and that grated. "I know a few that went back to academia," said Gindroz. "They felt that they were being pushed so much for the profit motive that their intellectual freedom was being infringed. That's almost exactly how one of them put it. They did not want to be con-

cerned about profits. They were more concerned about the development of science, than business."

Attitudes changed slowly. Gindroz, for example, oversaw patents.
One day, at a staff meeting, forms were handed out for collaborators
to sign giving the rights to the research to the company: "About half
opposed signing anything. They thought it was an infringement on
their intellectual freedom. Six months later, we did it again. The number in favor had changed dramatically. The attitude then was: Why
should we [the company] give away what we've worked so hard for?"

While Ashley tried to nudge Genetic Systems toward a more commercial focus, events began to overtake him. He found himself moving
from one crisis to the next. Financially, the situation was quickly deteriorating. The Legionnaires' test finally came to the market; orders
trickled in over the telephone.

> We didn't do phenomenally well, we did okay [said Ashley]. It was
> slowly building. Maybe it would get to four, five hundred thousand
> a year. Not bad. But it was never going to be *big*. You needed a
> core thing. That was sexually transmitted diseases. But that had
> been given away to Syntex. If you look at infectious diseases as a
> market, you discover that sexually transmitted is a subset, but a
> big subset—say 70 percent. You throw that part of the market
> away, you have 30 percent left. The sexually transmitted diseases
> could have carried the cost of the manufacturing and marketing.
> Now you have got to make all the costs on the rest of it.

The company struggled to market the infectious disease tests effectively without the lever of the sexually transmitted diseases. "Internally
we were focusing on the right problem," said Ashley. "And that's why
we ended up going after AIDS. In the beginning there was no reason
to believe we could beat Abbott. But they made a couple of fundamental mistakes, and our test proved to be better than theirs." By May,
Genetic Systems had gotten its test approved in Australia.

Technological excellence, however, was only part of the issue. Genetic Systems had two advantages: a good, basic test and sole rights to
the Pasteur patent. If, as Nowinski believed, Pasteur won its patent
suit, Genetic Systems could control the American market.[4]

But Genetic Systems labored under disadvantages that made a mockery of the forecasts Nowinski had offered up just a few months before.
At the Cable, Howse meeting, Nowinski had predicted approval by

March 1985. Clinical testing, however, did not begin until February, and FDA submission until mid-May; it was not approved until February 1986.[5] Meanwhile, in March, with Chemware still scattered across a lab bench in Seattle, Abbott steamrollered out the first AIDS antibody test. This test, hurriedly assembled and beset by false positives, ran on the Quantum, which the blood banks had long used in hepatitis testing. Finally, the Pasteur connection might eventually insure market dominance, but in the short term it created difficulties at the FDA. Genetic Systems, after all, was the only candidate *not* paying royalties to the federal government. "We were viewed," remarked Ashley, "as renegades."

The issue may have been moot. A year or so after approval, Abbott, with 70 percent of the market, was taking in $80 million in sales from its AIDS test. Abbott's test was highly profitable because it had already paid for the instrument, the plant, and the marketing team.[6] Genetic Systems had paid for nothing; all its big costs stretched before it. Second, the bulk of those profits—if they materialized—would be siphoned off: Pasteur would get half; the limited partners 12 percent; the marketer, American Hospital Supply, say 30 percent; and Applied Biosystems a chunk when, or if, Chemware saw the light of day.[7] Indeed, internal calculations showed that Genetic Systems would be lucky to come out with anything at all once the subtraction began.

And there were other problems to face: early on at least, the AIDS test sold almost exclusively to blood banks; American Hospital Supply was a powerful marketer, but *not* necessarily to the blood banks. Then, in 1984, American Hospital sailed into the same stormy weather that had buffeted Syva, Beckman, and other diagnostics companies. Cost containment devastated its core hospital business. After an aborted merger with Hospital Corporation of America, American Hospital Supply was swallowed up by its major competitor, Baxter Travenol. By the second quarter of 1985 it had pulled out of the deal. So a new marketing partner had to be found.

These difficulties drove Nowinski to another spasm of deal making. Once more he sought to save the day with a new relationship. To break free from the limits that Ashley, and the board, were building around him, Nowinski turned to Bert Bowers at Syntex, despite the increasing evidence—now openly discussed internally—that the Syva deal on the sexually transmitted disease tests was strangling Genetic Systems. Nowinski boasted of his close relationship with Bowers, and he speculated openly about the possibility that he might actually run Syntex one day.

GENE DREAMS

Bowers had his own problems: despite massive cost cutting, Syva continued to lose money. It desperately needed new technology, new products that would spread the risk from its now-outmoded systems. Nowinski thus found fertile ground when he went looking for a new marketing partner on AIDS. Nowinski began with the proposed deal the Genetic Systems board had approved: Syva to replace American Hospital as a marketer of the AIDS tests. The discussions soon took on a new tack, and Nowinski returned to Seattle with a much different deal. The differences may have seemed small, but the implications were profound. Nowinski had given Syntex the option *to buy* the entire diagnostics part of the company.

The deal shocked Ashley and the senior staff. Nowinski had not only free-lanced again, he had created vast technical problems for them all, and then, he had all but given away the only solid business Genetic Systems had. He had, in a sense, given Ashley's business away. Like many Nowinski deals, this one was a whirligig, full of options and moving parts. It was difficult to tell what was up; it was open to a variety of interpretations. Most immediately, Syntex agreed to buy 4.7 million shares at $8.50 a share for $40 million, or 18 percent of the company. Second, Syva would pay Genetic Systems another $20 million over five years to support diagnostics research. Third, Syva would market all AIDS tissue-typing products and, redundantly, the sexually transmitted disease tests. Fourth, Syntex would have the option to purchase the diagnostic operations of Genetic Systems at any time between August 1989 and July 1992. There was only one hitch: the deal could not be ratified until shareholders could vote. Thus, it had to wait until February and the annual meeting.

The Syntex deal seemed to embody the underlying strategy: use diagnostics to leverage into therapeutics. "We've always had as a strategic plan, the launch of the company through diagnostics, then a pharmaceutical approach for the long run," he said a few months later.

> That's what is happening here. We're bringing 60 million dollars into the corporation. After the Syntex transaction, we'll have in excess of 80 million dollars. We've got a very strong partner in diagnostics; we've got expanded market rights. So I think it's a very positive development. Besides, we don't know if they'll choose to buy diagnostics. Options are options. In a sense, they've already chosen *not* to buy diagnostics. I think Syntex has made a wise approach. Invest in us to obtain technology, and to look four to seven years from now to see if the two organizations are compatible. And whether they'd be better off leaving them alone . . .[8]

172

Permanent Revolution

Ashley had a different view. "It stunk. It was a disaster," he said. "He set us up with Syntex, and money was coming in, but they owned the company—the diagnostics company. By that point, I was beginning to give up." Nowinski's credibility sank again. All his talk about the bright future hung in the air—a mockery. And yet, despite the losses and the low stock price, Genetic Systems was not going broke; it could operate with its current cash for years, particularly if Ashley tightened management. What did Nowinski have in mind? Was he that desperate? Or was he shoring up his own position?

He seemed to be admitting that Genetic Systems could never make diagnostics profitable alone. This suggestion grew more plausible the closer one examined the deal itself. Nowinski and Bowers had sketched out the deal; actually implementing it was so complex that lawyers had to sort it all out. Syntex, with its legions of attorneys, wrote the first draft; Genetic Systems' lawyers made corrections. Negotiations began. There were large areas of ambiguity. Would Genetic Systems, as Nowinski argued, be able to comarket the sexually transmitted disease tests? Probably, but the return was uncertain. How much power would Syntex have to determine the direction of Genetic Systems research and new-product introductions? Quite a bit, it seemed. Why shouldn't Syntex do its best to depress the value of Genetic Systems by reducing its marketing efforts, just prior to the time it could buy it out? No answer to that one.

To Ashley, Genetic Systems was losing its substance. Nonetheless, Nowinski fought hard for the deal, and reluctantly, after much argument and wrangling over fine points, the board approved it, although it still had to go before the shareholders in February 1986. Two days later, the limited partnership closed, bringing in only $17 million, less than half the $35 million proposed in May 1984.

Soon Nowinski had another deal cooking. This one further undermined the talk of leveraging diagnostics into therapeutics—or, in fact, ever making it alone. It reinforced a sense among senior management that Nowinski had never really cared about building much more than a well-funded research boutique. For he now began talking privately about a way to bring in even more cash: a joint venture on therapeutic products.

CHAPTER 18

Buyout

Syntex has been a tremendous asset to us. It's been a very, very successful relationship. We've been able to develop our disciplines internally—research, development, marketing, manufacturing— while working side by side with a very experienced group. We brought to the story an extraordinary technology. They brought to the story the knowledge and the finances. We're going to keep parlaying that, but I think Genetic Systems is becoming an increasingly more integrated company.
                                                        —ROBERT NOWINSKI
                                                    interview, October 17, 1985

THE LARGE drug companies were engaged in a complex strategic game. On one level they still struggled to get a clear idea of the time frame for biotechnology. Would biotechnology become a major commercial factor in the late 1980s or the late 1990s—or never? Would there be limitations on profits, say, from problems patenting these large, complex natural macromolecules? How they felt on these questions—and the companies displayed a wide variety of opinions—determined further action. Second came the financial analysis. With biotech stocks falling in 1985, and with operational difficulties mounting, when would it be a good time to buy in cheaply? And would values ever become so cheap again? Third, what was the competitive outlook, that is, the relationship of one drug company to the other? The drug companies tended to see biotechnology— so small, so relatively underfunded—as a pawn in larger pharmaceutical wars. If one believed that antibodies were important, or developing in

174

a practical direction more quickly than other biotech products, there were only three major biotech players: Hybritech, Centocor, and Genetic Systems.[1] If one was taken out, that would increase the value of the other two and raise the possibility of being squeezed out.

By mid-1985, these factors had spurred a number of drug companies to action. Eli Lilly, for instance, already had quite a bit of experience in biotechnology. By 1985 it was actually selling Humulin, or human insulin, which it licensed from Genentech. And because its major product line, oral and injectable antibiotics, were made, like most biologicals, through fermentation, some cross-fertilization loomed. Lilly had also inaugurated on its own a major effort in cancer research with a strong focus on monoclonals. Still, Lilly had opened discussions with Hybritech over the last few years; it was, however, just one of a number of pharmaceutical companies "that came by to kick the tires," according to Hybritech's chief executive, Ted Greene.[2] But by 1985, Lilly was under increasing pressure: its main antibiotic product line was aging and facing pricing pressures worldwide. And just a few years earlier, a major new anti-inflammatory drug called Oraflex had killed several patients. Not only was Lilly deeply embarrassed—indeed, a few managers were eventually charged with failing to report early evidence of trouble—but the loss left a gaping hole in their product line.

Thus, Lilly fulfilled the criteria: it believed in the short-term possibilities of monoclonals; it had a need for new product and the cash to spend; and prices were at historic—if you think of five years as historic—lows. There was only one complication: Hybritech, or whomever Lilly bought, would have to want to sell out. Most of the large drug companies saw the futility of acquiring a company only to have all the important assets—the researchers—flee.

Still, Hybritech seemed a good match with large, venerable, and deeply conservative Lilly. While Genetic Systems rocked with personal, and strategic, disputes, Hybritech, perched on the California coast, seemed to cruise along. Brook Byers, venture capitalist, and Greene had a plan and stuck to it. Deal making was less a feverish dance than the unfolding of a design. They sprang few surprises on analysts. Greene's forecasts for revenue growth turned out pretty much as he had predicted: the first tests in 1981, the first profits three years later. Although Hybritech did not cure cancer—the company planned that as an encore—Wall Street retained its affections. Hybritech did have a three-year lead over Genetic Systems operationally, a fact Nowinski often brought up, but that did not alter the fact that as a public phenomenon, it was younger than Genetic Systems. It had waited patiently; it had been seasoned during its youth, like a minor league ball player, beyond the glare of public markets; it had laid down a foundation. Hybritech thus came closest to imitating that archetypal computer company, rolling out products soon after its offering.

It was, in fact, the exception that proved the rule. Ted Greene struggled to blend the systems of a large company with the entrepreneurial verve of the small—and in many ways he succeeded. Hybritech's early products astonished no one because Greene was after sales, not prizes. Marketing people liked them because they could sell them, but scientists shrugged. The tests employed monoclonal antibodies to detect certain large, well-understood molecules: immunoglobin E for allergy, HCG for pregnancy, PAP for prostate cancer, and ferritin for anemia. The most striking thing about them technologically was the format, the patented mechanism within the kits, the way of attaching radioactive tags to antibodies which made them work faster and more simply than conventional techniques.

Now talking about formats is like discussing a washing machine motor—necessary, but not about to make the pages of *Time*. Formatting was another in a long line of underappreciated technological assets: manufacturing, marketing, chemotherapy, linkers, polyclonals, diagnostics, *formats*. A format was the epitome of application. Finding the perfect monoclonal would not matter all that much if you used it in a mediocre format. Tandem—the trade name for this early series of tests—was thus a triumph of the tinkerer's craft.[3] It was a neat, workable invention. You could not call it a breakthrough or a blockbuster, although of course the company tried; it was not a revolution, although other companies borrowed it, and it swept away earlier formats. No one nominated Dr. Gary David, its inventor, for a Nobel Prize. All Tandem did was provide Hybritech with something credible to sell, allow the company to become acquainted with the ways of the FDA and iron out manufacturing problems before cranking out something *really* big.

Hybritech, like Genetic Systems, talked a lot about integration. That, of course, was the dream. Integration however, is a rhetorically flexible concept. Hybritech was somewhat closer to achieving true integration than Genetic Systems. It actually had built its own marketing group, forty strong by the end of 1984, although—and this was the way this kind of integration worked—its products were still sold through distributors.[4] And although it did its own manufacturing, it had to buy its instrumentation, its analyzers, from an outside vendor. And although it had a twenty-four-hour service hotline, it employed no service people of its own—integration, sort of. Still, Greene *had* played the game skill-fully. He avoided making near-lethal mistakes like the sexually trans-mitted disease deal or burning through his equity. He was on the road to building a company rather than a research boutique. Revenues had more than doubled from $13 million in 1983 to over $30 million a year later. *Product* revenues, which Genetic Systems never listed, rose from about $7 million in 1983 to $14.6 million, in large part because of a five-minute pregnancy test called the Immunoconcentrator, Icon for short, sold to hospitals and clinical labs. That resulted in a ten-cents-a-share profit, just over a million dollars—impressive, even if the profit came mostly from interest income.

And yet, for all of that, the waters were still treacherous. Abbott continued to gobble up market share, and Operation Neutral forced Hybritech to spend for its own analyzers. With less than one percent of a $5 billion industry, Hybritech was vulnerable. Product life cycles whirled past. To keep up, Hybritech had to pour funding into diag-nostics development. There were also patent disputes and lawsuits. Margins came under pressure. Meanwhile, Abbott, among others, readied a counterattack against Icon in the hospital market. And what of therapeutics? So far, Hybritech had never had to turn basic research into a therapeutic. Its diagnostic kits, for all their elegance, were a shuffling step, not a great leap forward. Making it in cancer therapy would require more money, perhaps $100 million, and new skills. Most of all it needed time. The cancer diagnostic kits—for PAP, CEA, alpha fetoprotein, and prostate-specific antigen—were still not all that spe-cific. They were for monitoring, not early detection. Nobody had yet found a cancer marker that really worked. The theory that diagnostic monoclonals against cancer would lead gently and easily to therapeu-tics—would, in other words, allow one to leverage diagnostics into ther-apeutics—began to look less likely.[5] For all its virtues—and there were many—Hybritech needed help. Good management made Hybritech more viable, but it was still probably not enough to ensure true drug-company-style integration. With the company making money, the tim-ing looked right to cash out.[6]

By early autumn 1985, management at Genetic Systems felt trapped and harried. Ashley's revamping was just taking effect. The AIDS test had gone to the FDA, and the infectious disease tests were in clinical trials. Chemware development wheezed on. A limited partnership and the Syntex deal—two very complicated arrangements—had to be cleaned up. The Syntex deal solved nothing. Personalities, rubbed raw by seventy-hour weeks and differences in fundamentals such as style and philosophy, banged together like hammers on sheet steel. A sort of shuddering malaise descended. To top it off, the irrepressible Nowinski began promoting his plan for the therapeutics program. Bristol-Myers had made a tentative approach a few months earlier, saying that it would eventually get back with "some thoughts on an offer." With that in mind, Nowinski presented a proposal to the board: a joint venture on therapy with Bristol, an exchange of technology for cash.

Ashley had had it. He felt that Nowinski had bludgeoned through the Syntex deal, and Ashley had sworn that it would not happen again. This was not the way Ashley thought professionals operated. As Nowinski gave his proposal to the board—"lecturing," Ashley said, "pen in hand, like the professor"—he jumped up. "Bob, sit down, you've said enough. Now I'll talk."[7] Ashley described other options, including selling the company outright. The dread words had now been spoken. After a discussion, the board told the two of them to come up with some sort of proposal—together. For the next three weeks, Ashley and Nowinski worked, locked together, like two nervous cats. It was uncomfortably tense. Nowinski banged away for a joint venture, a plan not unlike the original Oncogen agreement, and a paradigm of the career of any scientific entrepreneur: brains for cash. Genetic Systems would exchange its technology for $40 million. Ashley, not surprisingly, viewed such a venture as Nowinski's attempt to retain his own power. He would get millions more for research, but there would be nothing left of this *commercial* enterprise, Genetic Systems. All flexibility would be gone. Syntex had de facto rights to diagnostics; Bristol would get everything else. Another deal like Syntex would add so many operating complexities that the company would be bound as tight as a mummy. And it would put a cap on the stock. If you were going to sell it, why not really unload it and give shareholders something?

To dissuade him, Ashley attacked Nowinski's valuation as much too low. He argued that the technology was worth at least $140 million—hoping to set the figure high enough that Bristol would never go for it, and thus force a buy out. They argued for days. Finally, to settle the dispute, Nowinski assigned Udo Henseler, the financial chief, to come up with his own valuations. Three days later, Henseler came back with three different figures. Says Ashley:

> We ended up in my office. Bob and I are arguing back and forth and I can see Udo's head bouncing between us. Udo gives him the three valuations. Bob picks up the first one and puts it down. Picks up the second, and puts it down. Those were the high valuations.

Buyout

Then, he picks up the third one which appears to be lower. He starts talking off that one. So I say, "Look Bob, if I don't like this deal, it's not going anywhere," and I grab the third sheet. "Let me see that," I say. I look at it and it's just not right. That's when Udo pipes up and says, "I've been trying to tell you, I haven't finished that one." All that Nowinski did was look at the bottom line.

The two finally reached a compromise. The valuation would be set at $110 million; and although the pretext would be a joint venture, the goal would be a buy-out. Tensions eased. The board agreed that the two of them should explore the situation with Bristol. Meanwhile, the environment was changing, putting Genetic Systems, from Bristol's vantage point, in a new, attractive light. While Ashley and Nowinski argued, Ted Greene was deep in talks with Eli Lilly. On September 18, Lilly announced that it had reached an agreement to buy Hybritech.

Buy: the news rocked the industry. The drug companies were moving now. Somebody was actually willing to spend money for a biotech company. Maybe biotechnology was worth something after all. Lilly had agreed to pay twenty-nine dollars a share, or about $300 million, approximately 300 times Hybritech's 1984 earnings and 150 times its expected 1985 earnings. That was a 10 percent premium over the twenty-seven-dollar price when trading was stopped. Shareholders were to receive about twenty-two dollars in cash or notes, warrants for Lilly stock valued at four dollars each, and, as a sweetener, a new stock issue called a Contingency Payment Unit (CPU) which would be linked to Hybritech's performance. The CPU was designed for several purposes: to give Hybritech management a sense that it was still autonomous; to indicate to the markets that Lilly was not about to smother the entre-preneurial embers within Hybritech; and to offer Hybritech insiders a material incentive.[8] The CPU would open at three dollars a share and, if Hybritech proved profitable over the long run, could run up to twenty-two dollars a share by 1995, at which time Lilly would buy them back. Maximizing the CPU would add another $200 million to the deal.

The Lilly acquisition forced companies to consider the competitive question: with Hybritech gone, only two major public monoclonal com-panies remained. "I think a lot of pharmaceutical executives woke up after the Lilly acquisition and asked themselves not what is it going to cost them to be a player in biotechnology, but what is it going to cost them *not* to be a player," mused Peter Drake, then an analyst at Kidder Peabody.[9] Bristol-Myers, with its big cancer program, was one of them. Syntex had already made a deal for half of Genetic Systems; why could

179

Bristol not go after the other? What if Bristol should find itself locked out? What if monoclonals turned out to be the hottest thing in cancer since chemotherapy? The hazy therapy deal suddenly looked more attractive.

A team from Bristol, led by John Melton, the head of planning and development, flew to Seattle. The board had given Ashley and Nowinski permission to explore a buy-out, through the agency of a joint venture. Nowinski, with his tremendous gifts for the pitch, would present the proposal. Bristol made its offer. Nowinski and Ashley looked at each, and Nowinski stood up. "Let me suggest an approach," he said. He then went on to offer the absurdly high $110 million figure. Bristol then raised the possibility of a buy-out option, that is, the right to acquire Genetic Systems after a period of time, like the Syntex deal. Genetic Systems volleyed back: Sure, as long as *we* also get an option to buy you out first.

This, of course, was another absurdity. But it pushed Bristol closer to a straight acquisition. The Bristol team returned to New York to consider it. Several weeks later they called back to invite Ashley and Nowinski to New York. Negotiations now entered their final, serious phase. On Monday, October 21, Ashley flew to New York from Seattle; Nowinski had arrived earlier in the day after speaking to a blood-bank meeting in Florida. Although the negotiations were ostensibly secret, the stock ran up from $7 to $9.25 a share on Monday. Nowinski ordered the trading halted, showing his hand. The game was almost over now. On Wednesday the two parties met at nine o'clock at 345 Park Avenue. By one o'clock they had agreed to the following announcement: Bristol would buy Genetic Systems for $294 million, which came to about $10.50 a share. The shareholders would decide on the issue in February. That, of course, was when the Syntex agreement would be voted on as well.

Now the hard work began. The meetings shifted to the offices of Bristol's lawyers, Skadden Arps Meagher and Flom. Nowinski, Ashley, and Jim Lisbaaken, Genetic Systems' attorney, sat around a large table facing John Melton and his Bristol team, investment bankers, and lawyers. A section would be negotiated, then a group of lawyers would disappear to write it up. A dinner was held for the three Genetic Systems officers and the large group from Bristol on Wednesday evening. Then, they continued hammering out the final details. Finally, at one o'clock, Thursday morning, the final documents were signed.

Earlier in the week, Bristol had sent by messenger two thirty-page employment contracts to Nowinski and Ashley at the Intercontinental

Hotel. The two signed almost immediately, although Nowinski characteristically sought to fine tune his deal. Signing them quickly made sense to Bristol. The company was not about to spend $300 million to watch the two top people walk away. The terms were very generous: Stay five years and Ashley would get 20,000 free Bristol shares and Nowinski would get 30,000. Besides, tying up Nowinski and Ashley would deter Syntex from trying to get back into the game. Not that Syntex had much of a chance, although it did murmur a threat to sue, which precipitated another round of negotiations. Syntex eventually settled for a healthy profit.[10] Indeed, all the major shareholders were looking at windfalls: the Blechs would make over $10 million each; Nowinski $6 million; and Collinsen, the Allen group, and the New York investors millions each.

In Seattle, the buy-out came like a chilly, wet front off the sound. It was a rainy Monday, the temperature a dank fifty degrees. Five years ago, almost to the day, Genentech had gone public, but the parallel eluded anyone in Seattle. Gindroz, as the senior officer in Seattle—and as the investor-relations spokesman—knew what had happened in New York and called a staff meeting. Everyone filed into the auditorium, subdued, filling the seats, lining the walls.

"It was a deep shock—shock and hostility—to many people," said Gindroz. "The feeling was, why do we have to do this? I thought we were going to go on our own and we were going to get the benefit of all the great science we had developed. That's what Bob had been saying all these years. Many of the people felt as if a dirty trick had been played on them."[11] In the next few months, morale sank further. Nowinski returned from New York and explained that Bristol could afford to fund therapeutics, which was the real goal of the company anyway. In February, after the sale was approved by shareholders, Nowinski promised that no one would be laid off.

A week later workers in the first manufacturing facility were released. A few left; some found jobs elsewhere in the company. It could not be avoided, and it really was not his fault; nonetheless, Nowinski was blamed and morale sank further. It had happened too often before. Other layoffs, in areas that Bristol would handle such as public relations and finance, followed.

Wall Street debated the price. Had Bristol paid too much? No one could be sure. It was a question that could only be answered a few more years, or perhaps more, down the road. Opinions on how Genetic

Systems would fare within Bristol were flawed by the failure of analysts to know what really had been taking place inside the squat building on Puget Sound in the first place. If they were so wrong before, how could they predict after? Besides, $300 million was not a great amount of money for Bristol-Myers. It had $900 million in the bank, and it could undertake the acquisition without assuming much—if any—debt. It was an insurance policy, a preemptive strike. At the very least, it was not that Bristol feared Genetic Systems and Oncogen alone all that much; it was rather that it feared another drug company picking them up. There was talk that Bristol had panicked when Eli Lilly took out Hybritech and that it would have made a deal at any price. But that might have been a bit too strong. Whatever else it did, the Hybritech acquisition focused Bristol's attention.

The one immediate winner from the deal was Hybritech. As soon as the Genetic Systems buy-out was announced, Ted Greene began re-negotiating with Lilly. After all, here was Genetic Systems, unfocused, losing millions annually, getting the same deal as profitable—at least for the moment—Hybritech. A CPU is fine symbolism, but investors would like to see a bit more hard cash up front. If you were set on this figure, why didn't you buy *them*. Hybritech's pride was a bit singed. To complete the deal, Lilly came through with another $50 million, putting Hybritech out in front again.

The buy-out did not dampen the warfare within Genetic Systems; it simply altered the power equations. Nowinski was now in charge. Bristol believed in giving autonomy to its units, and Nowinski was their guy. He reported to the head of science and technology, an Italian-born executive named Abramo Virgilio.

In March, Ashley and Nowinski finally agreed on how to divide responsibilities. Ashley had resisted reporting to Bristol in New York through Nowinski; he wanted a direct line and he had threatened, despite the options, to resign. But Bristol balked and Nowinski promised not to interfere with diagnostics; he relented. The reorganization then proceeded. Ashley would take Genetic Systems and make it a pure diagnostics operation. The AIDS test had finally been approved. While Chemware languished, Ashley had his engineers building an instrument for doing both AIDS and hepatitis testing called Combo. Meanwhile, the Oncogen personnel working on cancer diagnostics were transferred to Ashley, burying the elusive idea of leveraging monoclonal diagnostics into therapy.

At the same time, Nowinski scribbled his own name atop Oncogen, which became a pure therapeutics organization. Todaro, in turn, found himself supplanted in the organization he had built. True, the acquisition brought some benefits. Bristol had plenty of research funds and a commitment to cancer research. And Bristol, arguably, had been more attracted to Oncogen—that is, Todaro and his team—than Genetic Systems. Todaro had to feel as if he had a key role in it all. All this, however, did not make his position more secure, although Nowinski did name Todaro scientific director for Oncogen and Genetic Systems.

At first, Todaro thought this might mean he would hold sway over both organizations. No way, Ashley told him; he refused to have his scientists directed by a research guy like Todaro. Ashley wanted to build products. He was not interested in what he viewed as pure research, so he would manage without Todaro. Ashley, who knew how to read an organization chart and liked Todaro personally, then asked: What is your role, George? Who reports to you? What real power do you have? Nowinski had told Todaro that he would act as an advisor, someone scientists could run ideas past. But, said Ashley, no one *reports* to you, and that's the key in corporate politics.

Todaro finally complained to Nowinski: Who *does* report to me? Finally, Nowinski redrew some lines: the patent attorney and the head of central services—the folks getting rid of toxic wastes, testing for radiation, mixing culture medium, and cleaning glassware—would report to Todaro. Everybody else at Oncogen would formally report to him. Soon after, Bristol ordered the patent attorney to report directly to New York. That left central services—the bottle washers.

Todaro had gone from the prestige of NCI, with his own lab that he ran as he wished, to overseeing the glassware cleaners in a distant colony of a huge corporation. Since leaving NCI, his old viral colleagues had scattered: Gallo had become famous because of AIDS; Aaronson was the name at NCI most closely tied to oncogenes; Phil Leder had a big lab at Harvard; Stephenson ran Oncogene Science; Scolnick was directing research at the biggest drug company in the world, Merck, with talk at the time that he might run it someday. There were consolations. In 1986, Todaro was elected by his scientific peers to the National Academy of Sciences, a much-sought-after accolade. That election was a sign of his deeper resources. He was a quietly stubborn man with methods of his own. He did not walk out; he had, after all, stock options and children going to college. He did not struggle as aggressively and openly as Ashley. He chose finally to ignore the whole political affair as if it had never happened, as if it were dream without substance.

GENE DREAMS

He pretended, like Nowinski himself, that organizational lines were just scrawls of ink on paper—meaningless, insubstantial. He still was the scientific star of the company; he had hired most of Oncogen's staff; he remained a formidable figure near the forefront of his field. Oncogenes, autocrine growth factors, oncostatins: These were breakthroughs that would follow him long after Genetic Systems was a footnote buried in an investment guide. In research, brain power was political power. By midsummer he was back redirecting the science as he had before that ominous word *buy-out* appeared. The clinical trials continued. The search for better monoclonals against tumors went on. Patents were filed. Even a mildly promising AIDS vaccine project began—an ironic return to viruses for Todaro.

Todaro's patience was rewarded. In 1987, Bristol brought Nowinski to New York. He was now given the large and impressive title of vice-president of new technology and chairman of Bristol's AIDS task force. Todaro remained in Seattle and was named president of Oncogen.

As for Ashley, within a year of the sale, his original fears came true. After a brief hiatus, Nowinski began to work his will within Genetic Systems once more. With Nowinski reporting to Virgilio, Ashley had nowhere to go except out. In December 1986 he resigned—part of a more general exodus—just before his AIDS project, the Combo, neared introduction.

Within biotechnology, the reaction to the buy-outs generated the immediate question: Who would be next? Rumors circulated. A third large drug company, Pfizer, was widely rumored to be stalking Genentech. The numbers were quite large. Genentech stock, at going prices, was worth over $1 billion, more when the limited partnerships were accounted for. Pfizer was a big company, but would it put out, say, $2 *billion*, without a guarantee that Genentech's most valuable assets, its scientists, would remain? Probably not. After all, unlike Hybritech and Genetic Systems, there was no reason for Bob Swanson to bail out now.

Meanwhile, Centocor, the major remaining independent antibody company, also came under scrutiny. Centocor had taken a somewhat different strategic approach from Genetic Systems and Hybritech. It had avoided the diagnostics wars by licensing its antibodies, often acquired from university labs, to larger companies such as Abbott. Centocor hoped to edge into therapeutics through antibody-based *in vivo* imaging systems, a sort of halfway house between chlamydia tests and cancer drugs. Its brain trust, Chairman Michael Wall and President

Buyout

Hubert Schoemaker, had longer horizons and perhaps a better sense of limits of finance and technology, although they too talked of making the leap to therapeutics. By the time Hybritech and Genetic Systems were seeking more permanent sources of money, Centocor was just coming into its own, although, like Hybritech, Centocor was certainly willing to entertain offers to cash out.

Ironically, the scent of takeovers began to bring investors back to biotechnology. This was the fulfillment of what some professional investors had been recommending for years: Focus on value, not operating results. The technology would bear fruit. Be patient. Wait. Companies will either make lots of money themselves, or be acquired at a premium. Wall Street thrives on takeover talk. Even if nothing happens—and in the case of Genentech and Centocor nothing did—takeover talk is good for business. Stocks trade hands, brokers start racking up commissions, investment bankers begin selling deals again, financial printers start putting in overtime, and journalists sniff around for new trends. Takeovers provide a jump-start for the market on cold winter days. So the conventional wisdom went, and so it was to be.

CHAPTER 19

The Iron Age

WHILE INVESTORS were cheerfully banking their profits from the two acquisitions, an ideology of technology and entrepreneurship was being undermined. Two shining entrepreneurial dreams had been absorbed by large corporations. Profitable investments for their early backers, the two companies had still not survived as operations. The reasons were complex, varied, and individual, and obscured by the air of congratulations. Although biotechnology had finally made some investors money—in the case of the Blechs and the early investors, a windfall—it was not about to drag American industry to a new capitalist millennium, not yet at least. While takeover talk pumped up stock prices, it also confirmed the harsh facts that Wall Street had intermittently acknowledged: the barriers to entry remained formidably high; the science, while promising, remained far more dependent upon serendipity than enthusiasts had proclaimed; and the expectations of Wall Street clashed, like the crosscurrents of some clanging atonal music, with the far longer cycles required by research, development, regulatory approval, and commercialization.

Entrepreneurs had a difficult year in 1985. As observers absorbed the mixed message of the two takeovers—and mixed was better than funereal—the microcomputer business, that incubator of Silicon Valley entrepreneurs, also fell on hard times. It had only been two years since *Time* proclaimed the personal computer its "Machine of the Year." Home computers had been the rage; computer companies, like their biotechnology cousins, had loaded up with cash; every town boasting a community development officer was trying to imitate Silicon Valley.

The Iron Age

The microprocessor, the computer on a chip, had successfully remade many of the products and processes of the Industrial West. And yet, computing, like biotechnology, was a mere collection of powerful *tools*, not a philosopher's stones. Home computers, as often as not, ended up stored back with the bowling ball and the CB radio. Computers did not create, from a vacuum, intelligence or wisdom; they did not, through their entrepreneurial agents, even have much immediate effect on that economic disease, anemic productivity. Even worse, Japan, a country not noted for entrepreneurs, was beginning to dominate the market for memory chips. Technology might incinerate the world, but it was not yet powerful enough to transform people's minds. The human material was far more conservative and resistant, far more *human*, than visionaries imagined.

The eclipse of that very model of the modern entrepreneur, Steven Jobs, also occurred in 1985. Ironically, Jobs was outmaneuvered by an executive from a giant corporation that made products superfluous to either personal growth or good health: a sugary beverage called Pepsi-Cola. Even more maddening, John Sculley, the former Pepsico marketing whiz who consigned Jobs to corporate limbo—a distant office, few responsibilities—proceeded to make Apple highly profitable again. Jobs retreated to reconsider his mission. He would turn his attention to education. Computers could, he argued, *revolutionize* education, and he would build a machine to do it. One had to grant him a certain flare for names, for the name of his new venture captured the indefatigable quality of the Silicon Valley entrepreneur: Next.

Jobs grasped the millennial expectations that clung to him. He assumed a tragic aspect: rejection, exile, return. The deep sense of frustration with the large corporation still existed. The fascination with technology's potential to transform and create did, as well. Jobs, or some other Jobsian figure, would return.

The rise of biotechnology altered, but did not transform, the basic forces at work in the traditional biomedical establishment. The pressures that produced the interferon hype still existed, indeed exacerbated by the talent and money going to the new companies. The difference now was that the companies and the academic community were bound more tightly together. An academic discovery could move stocks, particularly if there was a company directly involved; and, of course, the academic communities tapped the commercial money that was available.

Like the intermingling of foreign cultures, academia and commerce adopted a few of the characteristics of each other. The companies had their academic scientists, their campus environments, and the academic tendency to wander off in interesting research directions. Academia, on the other hand, became more responsive to commercial stimuli. The pure researchers had always drifted to the "hot" project, whether it was virology or cancer or oncogenes or immunotherapy, if only because that was where the money and the prizes flowed. The agenda, in turn, was set by their peers, who also acted as the prize givers, the paper reviewers, and the grant givers. By 1985 that had changed subtly. The institution of peer review remained, but a more complex interchange with biotechnology, including actual commercial ties, now supplemented it. The biotechnology companies might have only received grudging respect from the drug companies, but they had more influence on the academy. Projects that once had been considered hopelessly technological, or commercial, were now pursued regularly in the academy. It is, as yet, too early to tell whether that is good or bad for the long-term health of American biomedical research.

Interleukin-2 arose from that environment. As Mathilde Krim said in 1981, the cancer research community required a jolt of adrenalin now and again in order to give the public, and the large benefactors, hope. Indeed, all had been quiet since oncogenes peaked, say 1984; and even oncogenes, in all their complex splendor, lacked the simple narrative—*cancer cure*—of a mass enthusiasm like interferon. Now, in late summer 1985, as events at Genetic Systems were moving toward resolution, whispers began along the cancer research circuit about an NCI researcher named Dr. Steven Rosenberg, who, a few months earlier, had gained a measure of fame when he excised polyps from the colon of President Reagan. Rosenberg was working with a lymphokine, kin to interferon, called interleukin-2. Details were, as yet, sketchy. But gossip generates its own justification; interleukin-2 had an aura of something big, in part, because big people were talking about it. In late October the selection committee of the General Motors Cancer Research Awards met in New York City. Throughout the day, the group of distinguished researchers discussed candidates for awards.[1] That night they joined the press at a dinner at the restaurant atop the World Trade Center, high above New York harbor. At the table sat William Rukeyser, the editor of *Fortune* magazine. After dinner, various scientists arose to talk about new avenues of research; Rosenberg's results were described in radiant terms. Three weeks later, just prior to Rosenberg going public with his results in the *New England Journal of Medicine*, *Fortune* bannered his work as "a new cancer breakthrough" on its cover.[2] Researchers, said the magazine, have "now discovered how to use a small group of substances produced by the body's own immune system to control all cancers."

The interleukin-2 hype had begun. *Fortune* had made the splash; the rest of the media now tumbled furiously into the pool. Rosenberg, monkish in his white coat and spectacles, graced the cover of *Newsweek* and *Business Week*, popped up on the front pages of newspapers, and appeared on the "Today" show. Stocks began to move. Cetus, which had been rebuilt around anticancer lymphokines and which had supplied NCI with interleukin-2, soared from sixteen in mid-November to thirty-three, adding $387 million in market value. As in interferon's heyday, a scrambling pack of companies snapped at Cetus's heels. Even Genentech, which was not working on interleukin-2, but which responded with sympathetic sensitivity to any good biotech news, rose from forty-nine to the eighties by the first of the year. (It also helped that the first promising large-scale trials of tissue plasminogen activator were released in November 1985.)

Interferon now assumed a new place in the biotechnology firmament as a foreshadower of interleukin-2. Unfortunately, interleukin-2 then began to display its own imperfections. Rosenberg's technique turned out to be frightfully expensive and cumbersome. Moreover, Rosenberg worked with only twenty-five patients, each of whom received varying amounts of the protein, making it difficult to generate correlations between dose and response. Although tumors regressed in eleven cases, his regimen produced neither a cure nor conclusive proof of efficacy. And there were serious side effects: fever, chills, nausea, confusion, anemia—which required regular blood transfusions—and a high degree of fluid retention, which contributed to one death. Even if the trials had fewer difficulties, few onlookers considered the expense, the effort, and the time necessary to get interleukin-2 to the market, the crowded field working on it, the effect of a confused patent situation and of licensing deals or limited partnerships when they talked about the coming bonanza. Like interferon, the passion for interleukin-2 swept away all constraints. In their enthusiasm, even such magazines as *Business Week* and *Fortune* failed to delve into the problematic realities of immunomodulators such as interleukin-2 as potential products or of interleukin-2 as a *business.*

And yet the times *had* changed a bit: the cycles of hype and backlash were spinning faster. Less than a year later the biotechnology stocks were hammered with the news that interleukin-2 might not be a panacea after all. This time the media wielded the cudgel. A *Wall Street*

Journal story in October reported that tests were going poorly. Then in early December 1986, the *Journal of the American Medical Association* published a new paper from Rosenberg on interleukin-2 therapy, accompanied by an editorial from a Mayo Clinic physician attacking the procedure, a highly unusual public event that seemed to stem, at least in part, from the exaggerated uproar a year earlier. This time Cetus stock fell from almost twenty-four to nineteen before firming up again. Actually, if the market reacted too optimistically to good news, it responded too pessimistically to bad. As with interferon, the promise of interleukin-2 may lie in combination with other key proteins of the immune system, say a gamma interferon, a monoclonal antibody, or even a chemotherapy. The fact was, no one yet knew enough about the complex interplay of the immune system to know how interleukin-2 really worked. Interleukin-2, like interferon and most experimental cancer drugs, was a shot in the dark.

Nonetheless, by early 1986, some analysts were announcing a "new era" in biotechnology—this was at least the third "new era" in six years—characterized, once more, by the expectation of products. That, they argued, would make valuing these companies more rational than adding up doctorates or accepting the shifting value assigned by the market. Valuation, on Wall Street, had always been biotechnology's Achilles' heel. Valuation was an attempt to provide investors, notably the large institutions, with a rational framework within which to make investment choices. When was a stock fully valued, or overvalued? When should one buy or sell? *What* should one buy or sell? What was the fundamental value of the company as opposed to the epiphenomena of the stock?

In the past, analysts developed valuations by concocting a mix of factors: number of Ph.D.s, various market forecasts, burn rates, a gut feel. But with products so distant, one stood as good a chance of getting an accurate valuation from Wall Street as from a carnival fortune teller; perhaps worse, since fortune tellers presumably do not dabble in stock on the side. Analysts, even when offering their unbiased opinion, were victimized by an even more radical set of uncertainties than the market forecasters. Valuations teetered on changeable economic conditions, resting atop a shifting regulatory system which, in turn, covered a science in flux. Would the patent hold? Would that stuff work? Will management self-destruct? What will be the mood of the FDA? It was like sipping wine in an earthquake; with the glasses rattling, the table shaking, the rug shifting back and forth, and the very ground heaving, who could really tell the Petrus from the Haut Brion or the Chateau Plonk?

The buy-outs breathed new life into valuation efforts. If Genetic Systems was worth $300 million, what about Centocor or Genentech? Products seemed imminent. The nearer they came, the easier, in theory, to value them. A harvest of newly minted valuation models bloomed. Peter Drake, then an analyst with Kidder Peabody, made the first real splash when he came out with his Product Asset Valuation model.[3] Drake's model contained two parts: first, estimates of sales and earnings out to 1990; second, an attempt to determine the value of current assets in the future, based upon interest rates, risk factors and the beta, a value reflecting the volatility of a stock compared to the market, an exercise taught at most business schools. Drake produced detailed projections and balance sheets embellished with graphics. He estimated, for instance, that Genentech would be posting revenues from twelve products and three joint ventures in 1990, producing sales of $823 million; by the end of 1986 he raised his estimate on one product, tissue plasminogen activator, from $500 million to $800 million, gently boosting Genentech over Swanson's billion-dollar barrier.

There were other approaches. Linda Miller, an analyst at Paine-Webber, added up research expenditures and related them to the Genetic Systems and Hybritech acquisitions. This was a variant on the Ph.D. factor. She did not design this as a valuation model, although others soon used it that way. Bristol-Myers, for example, paid ten times the total amount of research money raised by Genetic Systems. Based on this, Centocor selling at twenty-four *should* be worth forty-eight. Stelios Papadopoulos, one of the newer crop of analysts—he was studying biophysics in 1980—offered up a system called the Biotechnology Technological Asset Valuation Model.[4] Papadopoulos, then at Donaldson, Lufkin & Jenrette, evaluated each company on a scale of 1 to 10 based on technology, management skills, and strategy, then compared them to each other. Genetic Systems got three 8s, Hybritech three 9s, Genentech three 10s. These subjective rankings were then combined with other factors—R&D income, expenses—to generate relative values. Papadopoulos was not determining stock prices, but values relative to each other; investors could, if they wished, use benchmarks like the Genetic Systems or Hybritech prices to orient themselves.

Each of these models attempted to quantify the unquantifiable. Despite the guise of rationality, all rested upon opinion—informed perhaps, but opinion nonetheless. Drake's attempt sat atop so many variables that it was like building on sand. His resolute optimism (The FDA would approve *twelve* Genentech products by 1990. Pretax margins would be 85 percent.) was perfectly timed for a resurgent bull market. But reality is cruel and small miscalculations, slippages inevitable even in conservative forecasting, tend to snowball. The R&D spending approach failed to account for the fact that some companies use research funds more efficiently than others. Genentech raised far more funds than either Genetic Systems or Hybritech. Did that mean that even at a billion, or two, in market capitalization, Genentech was undervalued? Papadopoulos exposed his own prejudices: trust me, trust my judgment. "The obvious caveat in employing this model is that it is *not* intended to be a black box for investment decisions," he wrote. "It is only a screen, a start, that directs the investor toward a group of potentially attractive investments."[5] There was more than a hint of realism here. Don't invest, he seemed to say, unless you already have a sense of what you're doing.

Papadopoulos, in particular, reflected on the difficulties of his role. This was unusual; reflection is not something Wall Street is known for. In March 1987, a company called Endotronics, which was attempting to commercialize cell-culturing techniques, collapsed. Papadopoulos took the occasion to reflect on the difficulties of biotechnology analysis.

> Biotechnology analysts may be of some use after all—no one recommended the [Endotronics] stock. The 1986 bull market induced most of the major Wall Street firms still holding on to buckle down and hire a biotech analyst. There may be twenty of us right now. . . . The complexity of the technology, the volatility of the stocks and the long-term nature of most endeavors undertaken by most biotech companies makes investing in this group a difficult proposition and begs the question of what the role of the sell-side analyst should be. Perhaps the most significant service we can provide investors is to keep them out of situations similar to the Endotronics one. Our personal approach has been to avoid what appears to be promotional play (that includes many companies with claims on AIDS cures) even if a small fraction of them turn out to be real.[6]

Papadopoulos was not alone. Experience *had* taught hard lessons. Promotion and hype continued, but without the unanimity of the early years. Realism and deflation shadowed hype more closely. Still, Drake's enthusiasm was shrewdly timed. With the market generally rising in early 1986, valuation models represented a symptom of an increasingly buoyant market, more than a cause. Investors needed a rationale to invest. In March a third wave of biotech offerings flooded the market. The boom lasted from early March, when Synergen went public, to June 18, when Cytogen, with Tom McKearn's linker technology, finally made it. More established biotech companies also rushed to raise money. Genentech continued to soar, driven by expectations for tissue-plasminogen activator (t-PA). Up into the nineties it flew in March, then split, then took flight again, before, a year later, splitting once more. By early 1987 it had repurchased its first two limited partnerships for five million shares, and it supported equity now worth over $3 *billion.* That put it within striking distance, in terms of equity, of established, integrated drug companies like Upjohn and Syntex. T-PA looked like the product that would make Genentech a billion-dollar company by 1990.

That now seems out of the question. Indeed, while Genentech did become the first biotechnology company to achieve true integration—it developed, manufactured, and sold t-PA by itself—it also experienced severe difficulties, some a function of overheated expectations, some limitations of the product and the marketplace. First, in May 1987 an FDA advisory panel failed to give Activase preliminary approval as analysts had expected. This delayed Genentech's entrance into the marketplace. Second, Genentech tried desperately to retain an effective exclusivity through a t-PA patent. However, it lost a patent suit and subsequent appeals against a British drug company, Wellcome, which was also developing a version of t-PA. Thus, although the FDA setback reduced the interval in which Genentech would be the sole supplier of t-PA, the failure in the patent suit insured that the global market would be more crowded and competitive. Third, and most ominously, another English drug maker, Beecham, introduced a competitive product, an altered form of streptokinase, called Eminase, in Europe. When t-PA, now called Activase, was finally approved in late 1987, Genentech's luck did not improve. After a rapid send-off, sales slowed dramatically as they approached the $200 million mark, a half to a fifth the range of earlier forecasts. A number of reasons were offered—physician confusion over clinical results, the high price ($2,200 a dose), the fact that it had to be infused intravenously—but the point here is that all forecasts, particularly in pharmaceuticals, are prone to error.[7] Genentech, in a sense, had been ambushed by the inherent conservatism of drug regulation and medical markets and by its own self-confidence.

Genentech found life as a pharmaceutical company far more difficult than as a biotechnology firm. Bob Swanson had indeed proven that one could turn an entrepreneurial biotech company into a full-fledged, integrated drug company. But he also discovered that the struggle only begins there. Having broken through, Genentech found itself in a kind of limbo, trying to compete with the large drug companies but with far fewer resources. No longer a company based on dreams, Genentech found that investors could begin to judge its prospects more realistically; not surprisingly, the stock skidded.[8] And the technology that Genentech, of all the biotechnology companies, proved so adept at exploiting did not provide the dominating edge that its promoters had predicted. Now Genentech faces life as an adult. To keep its stock price up, to avoid tempting an acquirer, the company must continue to generate more Activase-type products over the next decade, to continue to feed the entrepreneurial flames in an organization that will inevitably mature and bureaucratize.

Most of the rest of biotechnology could only dream of having Genentech's problems. Following the new-issue boom of 1985, harsh times descended again. Share prices were already sliding when the market crashed in October 1987; as in the past, biotechnology, with its tissue of expectations and its contingencies and unknowns, took the greatest beating. The investors that remained in the market, mostly large institutions, again fled to the fortresses of the blue-chip stocks. As 1987 became 1988, then 1989, little changed. Few new issues came public; one, a Seattle monoclonal-antibody company called NeoRx, had to first postpone, then reduce its offering. At the same time, other means of raising money, such as limited partnerships, were also affected: tax reform had made partnerships less attractive, and the capital they raised more dear. Even the venture capitalists felt the pinch. Without the ability to send companies public, they could not transform their investments into liquidity—into cash—unless they simply sold the company off; even in cases where offerings were made, as with NeoRx, they did not get their expected returns. While the entrepreneurial transmission belt did not grind to a complete halt, it did visibly slow.

Generally, the cost of money soared just when many biotechnology companies could least afford it. Many companies were finally pushing potential products through clinical trials and queueing up, like Russian housewives at the bread shop, before the FDA. Expenses were mounting; burn rates increasing; unexpected bills from endless patent disputes, troublesome manufacturing facilities, or FDA requests for more testing, piling up. Where could money be raised? The answer, foreshadowed by companies like Genetic Systems, turned out to be those dinosaurs, the big drug companies, which now danced through the blooming fields of biotechnology, plucking off licensing deals, setting up joint ventures, taking in the coming harvest. The big companies were willing to put up money, but as the skies darkened, they were able to extract more and more from the small. Almost no outright acquisitions as dramatic as Hybritech or Genetic Systems took place. Instead, the autonomy and independence, the ability to attempt that grand and glorious leap to integration, was gradually, piece by piece, compromised.

And so we enter what can only be called the iron age of biotechnology, where the struggle to survive, rather than triumph, predominates; where the reality of financing and technology may require companies to act as research boutiques rather than biopharmaceutical giants. Beneath the still-glittering surface, the substance of one company after another has been carted off. This dismemberment was rarely simple or obvious; one had to peer closely to see the range of tradeoffs involved. Take for instance a series of deals made by Nova Pharmaceuticals, another company assembled on the Genetic Systems model by the Blech brothers.[9] Nova focused on neuroscience and was led by Dr. Soloman Snyder, a leading light in the field, and based in Baltimore, near Johns Hopkins University, where Snyder had an academic appointment. Several months after the October crash, Nova closed a $42 million research and development partnership to support a brain cancer project. The partnership provided Nova with the money it needed to push the brain project forward; and it demonstrated that Nova could still generate support in the investment community (the Blechs by now were powerful, almost charismatic Wall Street figures). Finally, Nova was able to sweep the liability off the balance sheet, out of sight for the next few years. On the downside, partnership capital costs more than publically raised funds, and indeed, more than partnership money raised just a few years earlier. While the burden will not be felt for years, eventually investors will have to be paid, diluting either the equity base or the earnings stream. Still, that dilution was undoubtedly a gamble worth making; and the R&D partnership was something of a triumph for Nova, a testimony to the technology assembled by Snyder.

Several months later, Nova made an even larger, more complex and far-reaching deal. In May 1988 it announced a $49 million arrangement over three years with Philadelphia drugmaker, SmithKline Beckman. SmithKline agreed to buy 2.7 million shares, or 11 percent of Nova, for $25 million, at a slight premium to the market price. Tossed in were warrants to buy another 3 percent or 775,000 shares at $9.28 apiece. *And* SmithKline also won the option to pick up another $24 million in stock by 1991 at then-current market prices. By 1991 SmithKline could easily own more than 20 percent of Nova. Operationally, the keys to this deal are two joint ventures focusing on Nova's bradykinin antagonist—a biological that may mask pain by blocking the pain receptor in the brain—in cold medicines and ointments and in diseases of the central nervous system. Nova agreed to invest $49 million in the project over seven years for about 40 percent of the profits. SmithKline, in turn, will put several million dollars into R&D and $18.5 million into marketing and receive 60 percent of the profits. SmithKline marginally sweetened the deal by agreeing to include some of its traditional mental-health drugs in the joint venture and to allow Nova to screen chemicals in its drug library for activity.

Asking whether a deal like this is good or bad is almost beside the point. Nova needed cash to stay in business and it reportedly discussed first an outright acquisition with SmithKline, a possibility that was scotched when SmithKline began having difficulties of its own—its two largest drugs were getting hammered, its stock was sliding, ominous murmur of takeover talk had begun. Even with these problems, SmithKline still had the cash Nova needed, while Nova had developed the future products SmithKline required (in April 1989, SmithKline announced a merger with Beecham Group). A mutuality of perceived interests existed, and a deal was struck. Indeed, this kind of deal was widely applauded. Forty percent was quite a bit better than 5 percent— or nothing at all—and it was praised as representative of a new era when strategic alliances between biotech and drug companies were more equitable than in the past.[10] But there are other ways of looking at this as well. Nova, in all probability, would never have given away 60 percent of bradykinin if it had been able to raise money more cheaply; it will now take longer, all other factors being equal, to achieve critical mass. Meanwhile, if bradykinin flops, SmithKline can write off its investment. If it proves a great success, SmithKline retains the option to acquire. One can begin to hear the echo of Joe Ashley's concern after the big Genetic Systems deal with Syntex. Under these conditions, how independent will Nova now be to do other, similar deals? And how will the SmithKline deal affect future profitability? At what point does Nova cease to be an independent force and begin to act as a distant satellite of SmithKline?

And remember, Nova, like Genentech, is one of the lucky ones. It has cash, it has options, and if worse comes to worse, it has a likely acquisitor. What it lacks, and what the industry has gradually lost after the crash, was its autonomy, its freedom to act. Most companies have settled in as sort of halfway houses between the pure science of academia and the development and marketing of the drug giants. They act as research boutiques.[11] As in the iron age of Hesiod, the small must pay tribute to the large, the weak to the mighty. Capital is king. To break those chains would require a new generation of biotech products—and, of course, considerable skill and luck. As time goes on, as the techniques diffuse and mature, as the element of surprise is lost, the chances of that, of an insurrection, of a commercial revolution, grow more and more tenuous.

CHAPTER 20

Bumblebees and the Semiconductor Chip

ALMOST A DECADE has passed since Genentech's initial public offering. What has biotechnology brought forth? Widespread diffusion of the techniques: recombinant DNA, monoclonal antibodies. A handful of new products: some diagnostics that make possible quick and easy testing for pregnancy, ovulation, allergies, and a few bacterial diseases; a vaccine or two; therapeutics like alpha interferon that can shrink a few narrow classes of tumors, human growth hormone which can cause congenitally small children to grow, tissue-plasminogen activator which dissolves clots in arteries, and erythropoeitin for stimulating red-blood-cell growth in dialysis or cancer patients. Biotechnology General and Chiron race to commercialize superoxide dismutase for limiting the damage caused by clot busters like t-PA; Centocor, Xoma, Cytogen, NeoRx, and Immunomedics all have antibodies in clinical trials.

And what of cancer? Cancer had played such an important role in the market's early enthusiasm for biotechnology. Cancer had loomed behind interferon, behind magic bullets, behind oncogenes and interleukin-2. And so far, cancer had won just about every round. Despite a flood of new information, cancer has eluded definition. Cancer remains, at its heart, a mystery. Cancer has posed epistemological problems—for the market, for its victims, for the public at large. How do we interpret the facts? How do we know what we know? Cancer has taken the inherent difficulties of evaluating science-based companies and pushed them further.

Molecular biology argues very persuasively that it is unraveling the great clockworks of life. Depending on the point of view, this attitude is either a historical breakthrough or simple hubris, either one of the great advances in twentieth-century science or a plot by molecular biologists to dominate the academy and monopolize funding—or some combination. There is undoubtedly, in scientific historian Thomas Kuhn's phrase, a paradigm to molecular biology, a particular and distinctive way of viewing nature.[1] To the molecular biologist, the cell is a machine. If one can learn all there is to know about its mechanism, one can diagnose the problem and fix the cell.

This is an attractive notion. The key to understanding resides down among the molecules, particularly with that substance called DNA. Understand and control DNA, and one can possess a biological lever to move the natural world. Thus, interferon may not have provided a cure for cancer—not yet, one may hear from the back of the room—but it did demonstrate how genetic manipulation could produce large amounts of rare natural substances. And it focused attention on proteins of the immune system such as the interleukins which may, one day, contribute to that cure. Monoclonals may not be magic bullets, but they have many other uses, particularly in diagnostics, and they will certainly help to attack certain tumors. And oncogenes may not have been as clear-cut as first imagined, but the answers to the cancer puzzle, the molecular biologist would say, exist somewhere among the thickets of the gene. So give us the money and we'll find them.

A daunting task: With each new discovery, complexity seems to increase.[2] The genetic material alone, twisted tightly into the nucleus of the cell, contains billions of base pairs. An effort has lately arisen to obtain several billion dollars from the federal government to map the genome, to pin down every nucleotide sequence in its entire, winding way. A complex task—but the real complexity only begins there. Outside the nucleus, storms of cascading chemicals sweep through the murk: molecules couple and uncouple, determined by the complex behavior of atoms. Beyond the cell membrane surges a fluidic sea. Cells jostle symbiotically in massive urban populations, communicating, sharing resources, and coordinating to form organs and systems and, finally, organisms. Is DNA the magic key to this labryinth, or are there more profound interactions at other levels of organization? Does the control of this cell come from DNA managers alone or from some larger imperatives? The historian may ask: Does history lie with material forces, with great personalities, or with the march of ideas? So the biologist: Does a phenomenon like cancer begin in the DNA, in the interaction of cells, or in the atoms? Or is it more complicated than that?

Complexity is a subject Harry Rubin returns to again and again and again. To Rubin, one of the deadly scientific sins is to reduce complexity—to act reductionistically. In molecular biology, Harry Rubin passes for a pessimist. Rubin briefly appeared earlier as the instructor in the Cold Spring Harbor virology class that George Todaro attended in 1962. Today he is a professor at the University of California at Berkeley. Down the hall from Rubin works Peter Duesberg, a notable molecular biologist, who has become something of a naysayer on the great issues of the day, notably oncogenes and AIDS; he questions, for instance, whether HIV-1 is the cause of the disease.[3] Rubin has had his moments too. In 1980, Rubin wrote an editorial for the *Journal of the National Cancer Institute* questioning the somatic mutation thesis, the notion that damage to DNA is the primary event in carcinogenesis. This was like questioning whether Columbus was correct about the shape of the earth. Rubin voiced his objections, but then at the end of the paper, he suddenly opened up the discussion:

> Scientists seem to prefer questionable explanations to no explanation at all. In no field has this been truer than in cancer research, a veritable graveyard of once fashionable opinion. I find, on balance, the somatic mutation hypothesis to be at best inadequate. Yet there is no force that can resist an idea whose time has come, whether the idea be right, wrong, or simply inadequate. Timeliness, rather than the determination to take the whole evidence into account, may be responsible for the current popularity of the somatic mutation hypothesis.[4]

This is controversial stuff. But Rubin, with his long face and his white beard, keeps saying it. Rubin was trained as a veterinarian—he calls himself "just an old horse doctor"—in the days before molecular biology existed. There is, in his story, more than a hint of the moralist: he too once sinned. "I had all the illusions that I'm complaining about now," he says, in a low, raspy voice that drops now and then into aggrieved irony. "As a postdoc I studied basic science, particularly biochemistry. Eventually I came to Caltech, where I went in the reference lab and worked on the Rous sarcoma virus, trying to develop an assay for it." That assay resembled the one that so moved George Todaro at Cold Spring Harbor; in those days, it moved Rubin, too. In fact, Rubin used the virus, and the assay, to try to understand how normal cells transformed into cancer cells.

This turned out to be a problem.

The longer I went at it, the more I realized I would have to know a lot more about cells, to understand what the virus was doing to the cell. So I started studying cell growth regulation. The more I studied, the more complicated it became—the more I saw how naive we all were. We were deceiving ourselves by forgetting lots of things. We would see just a part of things. That would be okay as long as you got nice, well-defined data, and you didn't have to worry about all the related things. I had a sort of . . . obsession I guess, with trying to relate things that were going on in the whole organism.

Rubin found himself reveling in the model building of the molecular biologists:

That was some time ago. As you get more familiar with things, they are never as simple as they seem. You progress by making things simple, so they'll work. If you want to understand something as a chemical phenomenon of life, you eventually come up with a molecular explanation. But if you really want to understand the living phenomenon, you can get any explanation you want. So many things matter. Any time you take on a cell, it is impossible to know what is cause and effect.

Around this time, the Rous sarcoma virus got into the hands of real molecular biologists. They started juicing it; they became fascinated with exceptional things, with the same simplicity that I was drawn to in the first place. I could see people isolating these things away from reality, and convincing themselves and everyone else that it was reality. There's a sort of gentleman's agreement in science not to bring up such things. But it happens again and again. People build a grand conception from the results of very limited findings. And that raises everyone's hopes and expectations, whole fields get created, everybody gets excited: the grants givers, the prize givers, the newspapers, the science writers. It's

really a psychological phenomenon: everyone agrees not to disagree. And if you raise questions you get a very unpleasant con-
frontation.[5]

It was, then, an increasingly skeptical Rubin who first came across
work by a physicist and geophysicist named Walter Elsasser. A veteran
of the glory days of quantum physics, Elsasser fled Hitler in 1936 and
came to America. He won U.S. citizenship, served in the signal corps
during the war, and then entered academia. He began pondering biology, much as other physicists such as Niels Bohr, Erwin Schrödinger,
Max Delbruck, Leo Szilard, and even Francis Crick, had done; Elsasser
suggestively called his first book on the subject *The Physical Foundations of Biology*. Of him Rubin said:

Elsasser was putting into words, into philosophy, exactly what I
had discovered for myself. I had learned by watching things, working with them; he had a sort of mind that thought things out. I'm
no theorist. I had realized that biology was refusing to face the
search for real hard truths. And that molecular biologists refuse to
acknowledge that biology is even more subject to these basic questions than something as relatively simple as physics. Biology is very
empirical—I'm very empirical—but that doesn't mean you cannot
answer basic questions.

In physics, diverse observations are reducible to general formula
that cover different things. Biology is not like that. When it thinks
it is like that—for example, when the structure of DNA was discovered as a mechanism for replication—everyone goes ga-ga and
thinks everything is like that. It is very powerful, but quite limited
in its explanatory power. What Elsasser showed, and he knew com-
putation, is that if you had a computer as large as the universe
that could operate with a speed limited only by quantum mechanics, you could never completely figure out how a single cell oper-
ates. He also realized that some of the greatest concepts of physical
science are negative ones. The law of conservation of matter, that
you cannot get more energy out of a system than you put in. Or
that things can't travel faster than the speed of light. Or the uncertainty principle. These are limits. Biology today doesn't recognize limits. I found that once I made the break, I wasn't going to
be trapped into these restrictions to explain things by their components. My views as to what were acceptable changed. I was not
about to go tailing off trying to figure out the chemistry of some

202

interesting phenomenon because there are so many interesting connections. You have to maintain this sort of ambiguity, which can be unnerving. In that way it's like quantum mechanics: you can't picture energy being both a particle and a wave. As Bohr said, "No paradox, no progress."

Rubin took his doubts back to the lab bench. Instead of using specialized cell culture mediums—mixtures of nutrients and growth factors designed to make cells grow—Rubin decided to try to duplicate the *in vivo* environment of a mouse. He wanted to view the growth of tumor cells in a natural setting—or as natural as he could make it:

> I had these tumor cells that were misbehaving on me. I cloned and recloned and subcloned them, trying to fix their behavior. Well, I found that to be simply impossible. Tumor cells do not behave in any sensible manner, particularly when you transfer them from one environment to the other. They certainly do not fit any simple genetic determination. Every cell is different from every other cell—as Elsasser has said, every cell that ever existed is different from every other cell. That's the sort of paradox you find in physics. Not only is it wrong to think that the sole control of cell behavior begins with DNA, but to separate the cell from its environment, which is part of its heritability, is misleading. But you can say: With that attitude you will never learn anything. And you do have to have some of the reductionist spirit. But you have to realize that it's only part of the larger whole. It's a sort of schizophrenic existence: you work reductionistically, but you always have to be aware of that.
>
> Unfortunately, we've become slaves of our machines; the machines determine the way we think. You get all these machines, and you have to use them in the way that is appropriate for them. And it turns out that the most startling results can be gotten in very simple ways. In fact, not to do simple experiments is to deprive yourself of the chance and flexibility of looking at things as a whole. I sometimes think that if biology would cut back on the funds, it would be a healthy thing—except that the same people would get the money and peripheral types, like myself, would be cut off. I got cut off myself this year, but got it on a reconsideration. They said, "You're not going to find a mechanism." But I'm not looking for one. That's not what they want to hear.
>
> Science has become a religion and a faith. It has all the emo-

tional drag that goes with a faith. The trouble is, true religion is appropriate for revelation; science is not. I was standing in the hall one day, listening to a molecular biologist. He was Orthodox Jewish, and I heard him say, "There's nothing we can't do in biology." He shouldn't be thinking like that. Actually, it's a sin. Playing God. But people don't realize they've invested all their marbles in this game. It's become a matter of power as well as recognition.

Rubin clearly is an idealist, calling his colleagues back to the True Faith. He has a chilly eye; his cup is always half full. "The truth is, through all these enthusiasms—viruses, interferon, oncogenes—the cancer mortality statistics are simply not getting that much better.[6] That's not to say, we shouldn't work on cancer: It's formidable, tough, really interesting. But it's a disease of living. A part of aging. It's an illusion, or delusion, that you're going to be able to deal with the disease by popping a pill or taking interferon. It's probably a problem of the whole organism, and it's that kind of illusion that lies beneath the hype."

Sometimes he gets a bit depressed. "I was written off long ago," he says. "In a way, it's a favor. You don't mess around, waste your time in meetings and arguing and fighting." New companies do not beg him to serve on their advisory boards, and he has had those funding problems. Many consider him a nuisance or a crank—a latter-day Lamarckian chattering about passing on acquired traits—if they think of him at all. As one eminent scientist remarked about Rubin's notion that we can never know everything about a cell. "That's the bumblebee fallacy. You know, we can't quite figure out how a bumblebee can fly either. But who cares?"

Should Harry Rubin's gloom then be dismissed? Only at the risk of ignoring the uncertainty underlying basic biological questions.[7] For all the progress made over the last two decades in understanding cancer, the very basic arguments still rage: Is cancer one disease or many? Is it a unitary phenomenon—triggered by a simple common switch—or a heterogenous one? Does the answer lie in the genome, in some still-elusive pattern of oncogenes, or does it exist in some higher level of organization?[8] And cancer is not alone; fundamental questions rage about many intractable diseases, about basic questions of life, death, and aging. As much as molecular biology has learned about the immune system, for instance, more needs to be done before one can truly manipulate it.

Bumblebees and the Semiconductor Chip

This underlying scientific uncertainty creates significant commercial implications. From its earliest days, biotechnology has been continually offered up, particularly to the investment community, as the next great technological revolution, a worthy successor to the triumphant march of the semiconductor and computer revolutions. Such talk came not only from biotech entrepreneurs like Bob Nowinski, but from Wall Street brokers and analysts and from the press. Even industry critics, eager to indicate the depth of their concerns, would emphasize the unprecedented nature of the new biology—although they would often compare it to nuclear energy, with its weapons and fallout, rather than to the seemingly more benign microelectronics.[9] It was, like the magic bullet, an argument nearly impossible to bring back down to earth; like a cliché, it carried about itself the confidence of self-evident truth.

Actually, the notion of revolution could take on many meanings. They, in turn, could be sorted into two kinds of arguments. First, there was the long view that focused on the science. Biotechnology was portrayed as a chapter in a scientific revolution comparable to the impressive sweep of twentieth-century electronics, with its roots in the fertile soil of nineteenth-century physics. In both electronics and biology, a scientific revolution—"revolution" in the Kuhnian sense—spawns technologies that trigger the kind of "long-wave" economic revolution described by the late Austrian economist Joseph Schumpeter. The equation, simply put, has science creating technology, which, through the creative agency of the entrepreneur, then drives economic growth. The second view represented a popularization of the Schumpeterian thesis.[10] It focused more exclusively on the technological component and emanated strongly from Wall Street. It took as its model the invention, and subsequent development, of the semiconductor transistor. Given the obvious presence of a revolution—and in this popular version, science and technology tended to be confused—who would serve as the investable vanguard? The answer, borrowed again from the semiconductor experience, was an entrepreneurial company like Fairchild, National Semiconductor, or Intel (or in computers, Data General, Digital Equipment, or Apple).[11]

Are these analogies valid? It is difficult to gauge the profound changes that molecular biology may have on either medicine or, more speculatively, the economy over a period of decades. The essential question to ask here, however, is, Who shall control these technological changes that we have already witnessed? What will be the structure of the industry over the next decade? Why, unlike semiconductors and computers, haven't a wave of new companies swept aside the traditional drug companies?

Biology is far earlier in its development cycle than electronics. By the end of World War II, electronics had already been spinning off developments for over a century, beginning with the telegraph and telephone, followed by radio, radar, television, and the first crude digital computers.[12] In 1953, Watson and Crick unraveled the structure of DNA, five years after John Bardeen, William Shockley, and Walter Brattain developed the first semiconductor. The transistor was an application of science to construct a manufactured object, a technology; it led almost immediately to the first products and the first leapfrogging technologies. Watson and Crick worked from a much thinner scientific base; the structure of DNA was a paradigmatic event, but it did not spawn a technological breakthrough for twenty years. And even if one begins to count from Boyer and Cohen's 1973 recombinant experiment, the first product, human insulin, took over a decade to commercialize and was controlled, in the end, by a large company, Eli Lilly. While this is relatively rapid by traditional pharmaceutical standards—although these standards were changing as well over this period—it is, by the time frame of semiconductors, lethally slow.

Just as important is the relationship between the pace and structure of financing and product development. In microelectronics, the time it took to go from laboratory to marketplace was very short. The first commercial semiconductor product, a so-called point-contact transistor, went on the market in 1951, a year after the invention. By 1954 the number of firms producing transistors in the United States grew from four to twenty-six; by 1957 these new companies held about 64 percent of the market. Companies quickly became forces in the business. Texas Instruments had no electronics experience before 1949 and did not set up a laboratory until 1953. Nonetheless, it produced its first semiconductor in 1954; by 1960 or so it was a powerful force in the industry; by 1970 it would share in the glory of inventing the first integrated circuit. More remarkable was the rapid leap from organization to public offering. Intel, for instance, was formed by Robert Noyce and Gordon Moore, two veterans of Silicon Valley's first semiconductor firm, Shockley Semiconductor Laboratory, in 1968 (Noyce had been one of the inventors of the integrated circuit at Fairchild). Its first products, memory chips, appeared that same year. The company went public three years later, already a going concern. Public money was used for expansion, not creation.

Bumblebees and the Semiconductor Chip

Thus, the dynamics of the two industries were dramatically different. Although both semiconductors and biotechnology organized and tapped the public markets relatively quickly, the chip companies generated products either before going public, like Intel, or soon after. As a result, they were able to reduce almost immediately their dependence on outside financing. Because biotechnology companies went public so early in both their organizational and technological lives, they required a far longer nurturing from outside financing. Although dependence on outside financing is not necessarily lethal, particularly if prudently managed, the probability of failure rises quickly. A company dependent on outside financing is a company exposed.

Semiconductor history is characterized by rapid, continuous, and nearly exponential improvements on a basic technology. Science was important, but semiconductor companies were not waiting for conceptual breakthroughs on the frontiers of physics to generate new generations of products. Semiconductors, and Silicon Valley generally, were the province of engineers and inventors, not, for the most part, academicians and theoreticians. There was a strong consensus on fundamentals; there were no Harry Rubins in the chip business. New companies could steal the march quickly enough to beat back much larger, wealthier firms. Because they moved so quickly, money, particularly in the beginning, was far less important than technology.

In biotechnology, the translation of science to technology has, so far, not been as smooth. Companies were organized and built with Wall Street financing. Unable to strike quickly, they then had to turn to deal making—peddling off a project here, floating a partnership there—that made the kind of long-term profits needed for self-sustaining takeoff more and more difficult to generate and that made ever greater demands on the technology to create a breakthrough product. Biotechnology was not competing with the drug companies for financing; they were, to their long-term disadvantage, competing with other high-technology industries like computers. Although a few companies—Genentech, Amgen, and Centocor perhaps—may achieve that takeoff, the technology for the most part is being reabsorbed into traditional structures. Bristol-Myers buys Genetic Systems; Johnson & Johnson licenses dozens of products from dozens of companies; Merck pours hundreds of millions of dollars into a laboratory for molecular biology overseen by two former academics, Chairman Dr. Roy Vagelos and R&D chief, ex-NCIer Dr. Edward Scolnick.

Thus, although biotechnology has changed the way the drug business works, it has not sparked anything approaching a Schumpeterian economic revolution. Wall Street's willingness to raise billions in capital does not a revolution make; Wall Street annually throws millions away on unproductive investments. The major drug companies still dominate; despite generics and the increasing costs of R&D, the last few years have been some of the most profitable in history. So too, in its own way, despite its own problems, does NIH—witness the AIDS effort, which is essentially run out of Bethesda. At the same time, financing difficulties stemming in part from the stock market crash of 1987 have forced many biotech firms to license away major products to the large drug companies. The biotechnology industry increasingly looks like a middleman on a technological transmission belt between academia and pharmaceuticals. Meanwhile, the triangular relationship between academia, the drug industry, and the bureaucracy survives— although parts of the system have been battered, particularly NIH.[13] Indeed, the one move that might have radically altered those relationships—a 1988 Reagan Administration trial balloon to privatize NIH— was greeted by howls of protests.

How does the public fare in all this? We have seen the waste, the false expectations, the tendency of research efforts to go sniffing mindlessly after the hot new Wall Street trend, and the tendency to absorb some of the worst of the Wall Street and academic worlds and proclaim it wonderful. The public has been led hither and yon. But there is a more positive side to this as well. Although the myth of the scientific entrepreneur has proven less mighty than its promoters first suggested, biotechnology did bring a new electricity to a highly bureaucratic biomedical establishment. The drug companies were moving sluggishly in the late 1970s to absorb the tools of molecular biology, and academic researchers at NIH or the universities seemed to be settling for grand theory at the expense of practice. Biotechnology upset that complacency and forced those institutions to respond more quickly, more flexibly. Moreover, biotechnology—or whatever name it takes in the future—should fill an important role in those institutions if funding is available. The small companies can move more nimbly on thorny, if perhaps narrower, problems, than on large ones; they may also have more of a tendency to take risks, to tempt serendipity. The large companies, on the other hand, possess the resources to synthesize results into an encompassing systemic approach, particularly against chronic diseases such as cancer or those of aging. And the smaller companies can fill a social need as well, as a haven for researchers stifled by aca-

Bumblebees and the Semiconductor Chip

demia, repelled by large organizations, or interested in finally making a real buck.

And, perhaps, their day will come. The microelectronics analogy does offer the possibility that as the new biology matures, the true economic revolution may be yet to come. That will undoubtedly be held out for some yet as an incentive to form new, small companies, no matter the odds, and to provide the money for them.

CHAPTER 21

The Brightly Lit Stage

Being good in business is the most fascinating kind of art.
—ANDY WARHOL

APRIL 25, 1988: It is a warm, bright spring day in New York City. At the Sheraton Centre Hotel a two-day conference on progress in cancer research is being held. Bristol-Myers has two executives scheduled to speak: Dr. Stephen Carter, the senior vice-president for cancer research, and Dr. George Todaro, listed on the program as a vice-president of Genetic Systems in Seattle, but rumored to have now taken over the actual management of Oncogen. Todaro is slated to discuss tumor growth factor-beta and vaccinia growth factor, but at the last minute he has been delayed. Instead, Bob Nowinski, now a vice-president with Bristol in New York, steps in to take his place.

Carter speaks first. He casts a long shadow. He is, for one thing, a bear of a man; at cancer meetings he towers over the bald heads and the tweedy shoulders. For another, he represents Bristol-Myers to the cancer research establishment, and with Bristol spreading around awards and money, he has a certain clout and power. Carter is not an academic or a molecular biologist; he is a physician and a corporate executive. Until the early 1980s he had worked at NCI, refining chemotherapies, or as he refers to them, cytotoxic chemotherapies. He is an unabashed empiricist: mechanisms and theories do not interest him half as much as what happens in the clinic.

One day, a year or so earlier, Carter had spoken about biotechnology. "Some day it's going to be centrally important, though, personally I

The Brightly Lit Stage

think it has been oversold," he said, sitting in a large chair with Park Avenue, awash in Yellow cabs, receding through the window at his back.

It's not going to create the miracles that have been advertised. Five years ago, when I came here, it looked as if those miracles, with interferon and interleukin-2, were imminent. They were not only in and of themselves breakthroughs, but the cutting edge of a whole flood of products that were going to revolutionize the treatment of cancer, the business of cancer. But I think it was clear to many people, speaking privately, that a magic bullet was not going to happen and that interferon or interleukin-2 were not magic bullets. With luck, we might develop a new set of adjunctive treatments to standard therapies—surgery, radiation, chemotherapy, with all their imperfections. But that was not clear to laypeople. There are very strong pressures from oncopolitics and oncoeconomics to make biotechnology more miraculous than it really was. By oncopolitics I mean the need to keep the pump primed for cancer research; by oncoeconomics, the need to finance biotechnology companies. The two are synergistically linked.[1]

Thump. His elbows came down on the desk.

One of the things that strikes me when I read prospectuses or visit companies is the sense of unreality in their planning. It was all based on the magic bullet. They would lay out a scenario that would be absolutely true *if* they had the magic bullet. A drug so strong it would sell itself. The FDA so impressed that it would beg them to file an NDA with six pieces of paper—the fastest track in the history of medicine. I don't think any of the substances in the clinic are going to be important breakthrough products for the treatment of cancer. I doubt if most will be even commercial products. By breakthrough I mean something that obviates current treatment. Not just some $40 million product, but one large enough to build a large company.

That's the downside. The upside is that science is making tremendous progress in understanding the cancer process. We're at the point where the more we understand, the further we seem from practical progress. Again, that's because of the overoptimism over things like oncogenes which once seemed so straightforward. Now that we know so much more, it seems further away. There

211

is a critical mass developing, so that eventually certain synapses will be created allowing those intuitive leaps that result in real products. I personally think those sorts of things will happen around the turn of the century. At Bristol, we have to protect our current business, and prepare for the long-term breakthroughs, if and when they come. Ultimately chemotherapy will disappear. It's just a question of when.

You have to be absolutely crazy to believe that all of this basic science and understanding is not ultimately going to lead to important things. It's a tricky position I find myself in. I am almost a professional pessimist inside Bristol-Myers, because I feel I should bring reality to the current situation. And that reality is to dampen down oncopolitical and oncoeconomic hyperbole. But that does not mean that, in the longer term, I don't believe. Because I do. But I have made a wager—publicly—that in 1999 the major traditional forms of cancer treatment will still dominate.

Looming over the lectern at the Sheraton, Carter began his talk by repeating that same wager. His lecture set up a certain tension; he was undercutting the inherent optimism of a meeting called "Cancer Progress." The abstract of his paper in the program set the tone: "In cancer research today, there is an increasing gap between the implied promise of new findings and the reality of their ability to benefit patients today or within the next few years." And he proceeded to slam away. He mentioned the pervasiveness of oncoeconomics and oncopolitics; he listed ten examples of "oncopolitical hyperbole from 1970 to 1987." And he talked about the continuing promise of cytotoxic drugs.[2]

All that set the stage for Bob Nowinski, several speakers later. Carter and Nowinski would seem to have stood on opposite sides of a series of fault lines running through cancer research and through biomedicine generally. Carter was a clinician; Nowinski, in his day, an academic researcher. Carter represented an established drug company; Nowinski had been the charismatic head of an upstart, entrepreneurial biotechnology firm. Carter worked with and supported traditional chemotherapies; Nowinski pioneered biologicals. They were from different generations in the cancer war. Carter was based in New York City; Nowinski had been in Seattle. How had his view changed within Bristol-Myers? How would he reconcile himself to Carter's position?

Nowinski, slender and bearded, dressed in a baggy grayish suit—corporate by way of Soho—was introduced. He did not look all that much older than in the early days at Genetic Systems. With Todaro absent, he said, he would not speak in great detail about the tumor growth factors. Instead, he would take up the larger issue of the role of traditional pharmaceuticals versus biologicals. The major disadvantages of traditional pharmaceuticals were their low specificity, he said, hence their tendency toward side effects. "Drugs," he said, "are promiscuous because they are so small." Larger biologicals tend to be more specific and more potent. The disadvantage, unfortunately, is that biologicals are difficult to discover and produce, and almost none could be administered orally. But, he said, the differences between the two kinds of therapeutics is not as great as it would seem. Both operate in fundamentally the same way: by binding, in lock-and-key fashion, to a receptor.

It soon became clear that Nowinski was offering up a synthesis and laying down a new time frame. It would all develop much slower than either he, or Todaro, had been saying just a few years earlier.[3] In his scheme, biologicals would, in time, generate pharmaceuticals. The emphasis now was on *evolution*. The irrepressible optimism of the early 1980s had now been replaced with a more cautious, conservative view. Indeed, like Carter, Nowinski was now looking toward the year 2000 when orally administered biological compounds would begin coming onto the market in large numbers. Needless to say, Bristol-Myers had all the components for a biological strategy: some seemingly powerful antibodies against cancer; the oncostatins and tumor growth factors; an expertise with oncogenes; and of course, the most powerful group of cytotoxic drugs currently on the market. There was particular interest in tumor growth factor-beta, a family of proteins which not only seemed to accelerate wound healing but, a bit paradoxically, seemed to push cancer cells toward normalcy and even death. And there were the oncostatins—"like TGF-beta, a growth regulatory family, but far more diverse." Using these factors with other immunomodulators seemed to produce a synergistic effect. Using Oncostatin M with the highly toxic tumor necrosis factor has generated "an extraordinary amplification" in culture. "Synergy," Nowinski concluded, "is the way these things are going to occur."

The applause was polite, if not passionate. Nowinski answered a few desultory questions, then stepped down, and the attendees dispersed for the afternoon coffee break. Above the chatter and the rattle of cups, one was struck by the change in tone. This was a call to evolution, not revolution; to a long, difficult war one tumor type, one disease, at a time, not a sweeping victory; to difficulty, complexity, and uncertainty, not to powerfully elegant simplicity. This was no call to arms, no trumpeting of a revolution that would change the world tomorrow. The revolution clearly was dead—or it had not yet begun.

The urge to simplify and satisfy is a seductive one. It attracts the least sophisticated—those who find, say, Laetrile attractive—and the most advanced. Biotechnology sold itself on the belief that it could remedy the most profound economic and medical ills of the age, and please Wall Street as well. Such was the implicit pact made when Genentech sold its first shares of stock. The industry quickly assumed the coloration of its time: a bit blustery, a bit loud, with a nervous undercurrent. Talk of medical breakthroughs became pretexts to raise more money; capital accumulation was confused with speculation; rhetoric was mistaken for reality. Companies wielded complexity like a weapon: results so sketchy and ambiguous that they could be interpreted freely, fantastically. The sheer distance from lab to clinic, from cell culture to human patient, created a sort of imaginative space and nurtured dreams of miracles and money. And yet seven years later, the realities of drug development persist; there are no miracles, only developments.

Still, the taste for miracles has not abated; it lies embedded deeply within a society that has accepted the notion of health care for all and a culture bent on long life on this earth. It is a passion that blurs all distinctions. Genetic Systems, for all its flaws, was a solid scientific enterprise; Alfacell, on the other hand, has proven thus far to possess the substantiality of smoke. But both appeared—both were sold—on the same brightly lit stage, dressed in similar gaudy costumes. They were joined by scores of others. Most featured the same apparatus: advisory boards, mice, consultants, forecasts, and laboratory miracles. Many called themselves "biotechnology" companies. Distinguishing one from the other was as tricky as choosing an Andy Warhol soup can from a shelf full of imitators—or from the real thing. Like the art market, biotech posed complexities and uncertainties that created a role for experts as powerful and indispensable guides. Promotion and deal making followed. Values inflated as the ability of the language to make precise distinctions declined. No concept has undergone quite as much abuse as "revolution."[4] Revolution today means less the turn of the wheel or a bloody, radical change or a crash of a government than a new hat or a new dress or a fresh fashion in miracles. The action unfolds upon the surface, feeding a continuing social dream of transformation; of something new, young, and different; of a visionary to lead the way to the promised land. Thus, the vanguard—an entrepreneur, an artist,

214

or some Warholesque blend of the two, who possesses some secret knowledge, some gnosis, of the capitalist millennium.

Selling Genetic Systems or Alfacell was not that much different from selling, say, a painting by Warhol, Robert Rauschenberg, or Jasper Johns. The latter-day entrepreneur carried the aesthetic culture into the society of business.[5] Without products, distinctions become a matter of opinion. Substance or technique begin to matter less than ideas and style. It becomes a pop world, where anything is possible. Fortunately, beyond the selling, beyond the similar aura cast by these kinds of goods, in the marketplace of opinion, art and medicine *are* finally very different things. Art must wait for history to judge its merit; until then, the game belongs to dealers and critics. In biotechnology and medicine, theories must sooner or later prove themselves among the sick and dying—an unforgiving jury—and in the marketplace of goods. What the replicable experiment is to science, the marketplace is to technology. Does the tumor shrink? Are there side effects? Can the technology be sold profitably, or should it be put aside? How does it match up against that single objective benchmark, the therapeutic index, of traditional therapies? Does it provide much help to the physician? In that way, it finally gains exit from the world of fashion and hype.

NOTES

Introduction

1. Horace Freeland Judson, *The Eighth Day of Creation* (New York: Simon & Schuster, 1979).

2. See *Who Should Play God?* by Ted Howard and Jeremy Rifkin (New York: Dell, 1977). Howard and Rifkin paint a grim picture that wildly overestimates the short-term capabilities of genetic engineering. The authors took some of the wilder musings of academic scientists and turned them into flat predictions: that humans could be cloned in a decade or that massive eugenics programs were about to commence. The pair combine sensationalism with paranoia. "Today, only a handful of people are privy to the secret of life and how to manipulate and change it," they write. "It is now only a matter of years before biologists will be able to irreversibly change the evolutionary wisdom of billions of years with the creation of new plants, new animals and new forms of humans and post-human beings" (p. 8). Rifkin, of course, continued with other books and relatively effective campaigns to halt some agricultural biotech experiments.

3. Leo Marx, *The Pilot and the Passenger* (Oxford: Oxford University Press, 1988).

4. James Carey and John Quirk, "The Mythos of the Electronic Revolution," *The American Scholar* (Spring 1970): 220.

5. Gunther Stent, *The Coming of the Golden Age: A View to the End of Progress* (San Francisco: University of California Press, 1969).

6. See Marshall Berman, *All That Is Solid Melts into Air: The Experience of Modernity* (New York: Penguin, 1988), p. 81. Berman also notes how Stent's views were adopted by California governor Jerry Brown in the mid-1970s, and how one cannot today look back on them "without feeling nostalgic sadness, not so much for the hippies of yesterday as for the almost unanimous belief—shared by those upright citizens who most despised hippies—that a life of stable abundance, leisure and well-being was here to stay" (p. 82).

7. See Martin Kenney, *Biotechnology: The University-Industrial Complex* (New Haven: Yale University Press, 1986), pp. 2–3. Kenney refers to a further elaboration of the Schumpeterian thesis in his own unpublished manuscript.

8. Judson, *Eighth Day of Creation*, p. 637.

Chapter 1

1. See Stephen Hall, *Invisible Frontiers: The Race to Synthesize the Human Gene* (New York: Atlantic Monthly Press, 1987).

2. John Brooks, *Once in Golconda: A True Drama of Wall Street 1920–1938* (New York: Norton, 1969), title page. Brooks specifically referred to the years between the wars, particularly the 1920s, when Wall Street seemed to be a Golconda.

3. Sir Peter Medawar, "Induction and Intuition in Scientific Thought," from *Pluto's Republic* (New York: Oxford University Press, 1982), p. 73.

4. This, of course, is a generalization. There were exceptions, though they were few and far between. The 1988 Nobel Prize for medicine was given to three pharmaceutical researchers who pioneered the more inductive approach to drug design: Sir James Black,

Notes

who discovered the angina drug propranolol at ICI and the antiulcer cimetedine (trade name: Tagamet) at SmithKline Beckman; and Dr. Gertrude Elion and Dr. George Hitchings, who developed acyclovir, an antiherpes drug, among others at Burroughs Wellcome.

5. See Victoria Harden, *Inventing the NIH: Federal Biomedical Research Policy, 1887–1937* (Baltimore: Johns Hopkins University Press, 1986).

6. NCI budget figures come from the *NCI Fact Book 1981* (Washington, D.C.: NCI).

7. Horace Freeland Judson, *The Eighth Day of Creation* (New York: Simon & Schuster, 1979), p. 613.

Chapter 2

1. See William Broad and Nicholas Wade, *Betrayers of the Truth: Fraud and Deceit in the Halls of Science* (New York: Simon & Schuster, 1983). Broad and Wade strip away the scientific ideal and expose the somewhat less appealing, albeit more human, picture below.

2. Quoted in *Fortune*, Oct. 1, 1984, p. 9.

3. *Barron's*, Jan. 11, 1988, p. 68.

4. Quoted in Sharon and Kathleen McAuliffe, *Life for Sale* (New York: Coward, McCann & Geoghegan, 1981), p. 38.

5. *Fortune*, July 9, 1984, p. 140.

6. Farley was often quoted in this fashion. Quote in text comes from John Elkington, *Inside the Gene Factory: Inside the Genetic and Biotechnology Business Revolution* (New York: Carroll & Graf, 1985), p. 46. Also see *Fortune*, June 16, 1980, p. 149, where Farley says, "Our intention is to be the IBM of genetics."

7. Boyer never actually worked full-time at Genentech, although he did make paper millions when the stock went public. He acted, instead, as a sort of recruiter, éminence grise, and front man for Genentech. He proved, for instance, to be very adept at communicating science to the public. He was criticized in some circles for his opinions and accused of hyping technical advances. In time, Boyer stopped giving interviews and returned to the science, less an entrepreneur than a consulting entrepreneur. Although he was elected to the National Academy of Science in 1985, he still has not received a Nobel Prize, a failure many observers ascribe to his commercial activities.

8. That, in fact, was true. But there were charges of increasing secrecy not only about developments within the company, but with scientists consulting or under contract to Genentech. This was particularly the case with Boyer's UCSF lab. See William Boly, "The Gene Merchants," *California* (Sept. 1982); and Martin Kenney, *Biotechnology: The University-Industrial Complex* (New Haven: Yale University Press, 1986).

9. Randall Rothenberg, "Robert A. Swanson, Chief Genetic Officer," *Esquire*, Dec. 1984, p. 366.

10. As late as April 1986, Swanson appeared on the cover of *Business Week* in the white jacket, the red tie, and the lab in the background. The cover text summed up the Genentech view: "Biotech Superstar. Wall Street Loves Genentech. The reason: It's on its way to becoming a major pharmaceutical company."

11. Nelson Schneider, interview with author, Dec. 16, 1986.

Chapter 3

1. Cory SerVass, "Can Interferon Prevent Cancer?" *Saturday Evening Post*, March 1981, p. 76. Under its editor, Dr. Cory SerVass, the *Saturday Evening Post* has been a strong promoter of alternative approaches to cancer treatment over the years. But while her enthusiasm today seems overwrought, the treatment afforded interferon in more sober publications was not, qualitatively, much different.

2. Quoted in Sandra Panem, *The Interferon Crusade* (Washington, D.C.: Brookings Institution, 1984), p. 11. Panem's book is the best study of the interferon phenomenon.

Notes

3. "A. D. Lasker Dies; Philanthropist, 72," *New York Times*, May 31, 1952, p. 1.

4. See George Johnson, "Dr. Krim's Crusade," *New York Times Magazine*, Feb. 14, 1988, p. 30.

5. What constitutes a brilliant researcher? In a November 1985 interview with Krim, *Omni* magazine describes her as a "brilliant researcher" and recounts how, at 22, she used one of the first electron microscopes to actually "see" DNA for the first time in history. No mention of this breakthrough appears in other accounts of her life.

6. Lee Edson, "A Secret Weapon Called Immunology," *New York Times Magazine*, Feb. 17, 1974, p. 58. Immunology, or immunotherapy, remains an immensely appealing slogan in the popular culture. There is, for instance, Dr. Stuart Berger's best-selling book *Immune Power*, a dietary plan that purports to treat any number of maladies by bolstering the immune system through nutritional supplements. Berger, a skillful marketer, thus combines two unproven, though seductive, articles of modern faith: immunotherapy and diet.

7. Jan Vilcek, interview with author, May 2, 1986.

8. Quoted in Panem, *Interferon Crusade*, p. 19.

9. Quoted in Michael Edelhart and Jean Lindemann, *Interferon: New Hope for Cancer* (Reading, Mass.: Addison-Wesley, 1981), pp. 150–51.

10. See Panem, *Interferon Crusade*.

11. Quoted in Edelhart and Lindemann, *Interferon: New Hope*, p. 30.

12. Quoted in Panem, *Interferon Crusade*, p. 99.

13. In late 1988 the FDA approved alpha interferon for use against Kaposi's sarcoma, the cancer associated with AIDS. By that time the total market was about $75 million annually, with the expectation that the new indication, and others, such as its use against genital warts, would expand it further. Financially, then, alpha may over time prove to be a nice middle-level pharmaceutical—not a blockbuster, but not a loss leader either.

14. Gerald Weissmann, *They All Laughed at Christopher Columbus: Tales of Medicine and the Art of Discovery* (New York: Times Books, 1987), p. 70.

15. Jan Vilcek, interview with author, May 2, 1986.

Chapter 4

1. For two differing accounts in the business press, see Laura Saunders, "I Can Do That," *Forbes*, July 18, 1983, p. 36; and N. R. Kleinfeld, "Birth of a Health-Care Concern," *New York Times*, July 11, 1982, sec. D, p. 1. In the *Times* article the Blechs are actually described as looking for a "short-term hit . . . some aspect [of genetic engineering] that could put the product on the market in a few years and thus show investors what the end of the rainbow looked like." The *Forbes* article examined the Blechs' growing empire.

2. Quoted in Grant Fjermedal, *Magic Bullets* (New York: Macmillan, 1984), p. 6.

3. David Blech, interview with author, May 19, 1987.

4. Johnston would form Cytogen again with a different group of researchers. "No question," said Johnston in a 1987 interview, "that Nowinski was the guy to make it [Cytogen] go." When Cytogen collapsed, Johnston was angry. And although he was compensated sufficiently not to take it to court, he still viewed the affair for what might have been. "I guess you could say if we had allowed Cytogen to go to Seattle with Nowinski, the price of the company at the time of the sale would have been worth $25 million."

Chapter 5

1. For a general history of the founding, see Genetic Systems' initial offering prospectus, June 4, 1981.

2. Certificate of incorporation of Registrant, exhibit 3–01 in Genetic Systems' SEC filings.

3. Genetic Systems' initial offering prospectus, June 4, 1981, p. 33.

Notes

4. John Simon was one of those eager young men who Charlie Allen set loose. His father had worked for many years at Syntex as a scientist, and he had completed a degree in chemical engineering, supplemented by degrees in business and law. He would go on to serve as one of the central investment bankers for Genetic Systems.

5. In an interview, Simon remembered it slightly differently. Three-way negotiations were going on, but he thought Blair would do the deal even if the Allen/Syntex agreement wasn't made. It was just a question of price—which would vary depending on the Allen/Syntex agreement coming to fruition. "One didn't depend on the other, as much as help the other," he said. "We would have made the investment anyway, and D. H. Blair would have underwritten it" (interview with author, April 10, 1987).

6. James Glavin, interview with author, Feb. 1987. All further quotations from Glavin in this chapter are from this interview.

7. Herbert Allen bought 150,000 shares; Marvyn Carton, long a powerful Allen partner, particularly in the movie deals, picked up 175,000; and two investment operations, Wood River Capital and American Diversified Enterprises, took 375,000.

8. Promoting new issues could take on a carnival air. In 1985, Davis and Blair agreed to underwrite a company, backed by *Penthouse* magazine publisher Bob Guccione, developing commercial uses for fusion energy. I attended the due diligence meeting—the forum meant to inform potential investors about the company—at Guccione's twin townhouses off Madison Avenue in New York City. With food, liquor, a handful of Pets, Guccione, his dogs, partner Kathy Keeton, and lots of brokers, the presentations were soon drowned out in merrymaking. "Look," Davis finally cried. "If you want to live in a house like this, buy this stock." End of due diligence. The offering failed.

9. The press on Davis makes for lively reading. See Edward Boyer, "$250 Million Isn't Enough for Morty Davis," *Fortune*, April 16, 1984, p. 102; and the short bio on Davis that appears as part of *Financial World*'s "Wall Street 100," July 28, 1988, p. 48.

10. Genetic Systems' initial offering prospectus, June 4, 1981, p. 8.

11. John Simon, interview with author, April 10, 1987.

Chapter 6

1. Watson, ironically, was instrumental in altering the potent mythology of the molecular biologist through *The Double Helix* and arguably preparing the ground for the exodus of researchers into commerce. Not surprisingly, the book was greeted in some circles of the popular press and scientific community with outrage when it appeared in 1968. Gunther Stent has compiled a fascinating selection of reviews in a 1981 volume, *The Double Helix: A Critical Edition* (London: Weidenfeld & Nichols). The "Lucky Jim" came from Sir Peter Medawar's essay "Lucky Jim" in *Pluto's Republic* (New York: Oxford University Press, 1968), p. 270. Medawar, let it be said, argued that Watson was both very clever and very lucky.

2. Quoted in Edward J. Sylvester and Lynn C. Klotz, *The Gene Age: Genetic Engineering and the Next Industrial Revolution* (New York: Charles Scribner's Sons, 1983), p. 48. As an institution builder, Baltimore was the Watson of his generation: After his Nobel Prize, he went on to become the first director of his own major research laboratory, the Whitehead Institute on the MIT campus in Cambridge, backed by money from Edwin Whitehead, the founder of Technicon, a health-care company he sold to Revlon. Whitehead's bequest—financing for a 130,000 square foot facility, plus $5 million a year for operating expenses, $7.5 million to the MIT endowment, and $100 million after his death—was, as Martin Kenney has written, "the most important biological research institute created in the last fifty years" (*Biotechnology: The University-Industrial Complex* [New Haven: Yale University Press, 1986], p. 50).

3. Horace Freeland Judson, *The Eighth Day of Creation* (New York: Simon & Schuster, 1979), p. 67.

4. George Todaro, interview with author, March 13, 1986. All further quotations from Todaro in this chapter are from this interview.

5. Quoted in Natalie Angier, *Natural Obsessions: The Search for the Oncogene* (New York: Houghton Mifflin, 1988), p. 65.

Notes

6. A few viruses have been implicated in cancers, although they seem to be rare in the wide spectrum of the disease. For instance, Robert Gallo's laboratory at NIH has isolated several members of the HTLV—or human t-cell leukemia viruses—that seem to trigger leukemia. These viruses, of course, are cousins to the virus associated with AIDS. And indeed, AIDS can be viewed as the mirror image of leukemia: instead of a wild proliferation in white blood cells characteristic of leukemia, AIDS features a total shutdown of the immune function.

7. Philip Rahv, *Essays on Literature and Politics, 1932–1972* (New York: Houghton Mifflin, 1978), p. 6.

Chapter 7

1. George Todaro, interview with author, March 13, 1986. Unless otherwise noted, all further quotations from Todaro in this chapter are from this interview.

2. See Raymond Ruddon, *Cancer Biology*, 2d ed. (Oxford: Oxford University Press, 1987), pp. 244–73.

3. George J. Todaro, *Autocrine Secretion of Peptide Growth Factors by Tumor Cells*, National Cancer Institute Monograph 60 (Washington, D.C.: NCI, 1982), p. 146.

4. Ibid.

5. Ruddon, pp. 265–66.

6. *NCI Fact Book 1982* (Washington, D.C.: NCI), p. 15.

Chapter 8

1. Genetic Systems' initial offering prospectus, June 4, 1981.

2. Joseph E. Eichinger and Robert Kupor, "Genetic Systems Corporation," Cable, Howse & Ragen, Dec. 12, 1983, p. 9.

3. D. H. Blair & Co., "Genetic Systems and the New Age of Medicine," Jan. 12, 1982.

4. There were other caveats to all this, of course, but they tended to be downplayed or buried in the numbers. Chlamydia infections were often so mild that they were shrugged off as just another yeast infection, or so obvious that physicians would not feel the need to order a diagnostic. Herpes, a virus, could not be cultured at all, and since there was no cure for herpes, there was little benefit to a herpes diagnostic.

5. D. H. Blair, "Genetic Systems," pp. 2, 5.

6. Ibid, p. 2.

7. Robert Nowinski, interview with author, Oct. 17, 1985.

8. Howard E. Greene, Jr., "Hybritech Incorporated: Presentation to the Financial Community," Hybritech, Nov. 1982. Greene's use of Merck is a little off the point here. Merck did not grow solely on the power of antibiotics; a better example might have been Eli Lilly. On the other hand, Lilly had a quite respectable, if not as high-powered, business, which began in the late nineteenth century; antibiotics powered it to new heights but did not, by any means, establish it in the first place. Merck, too, had a long operating history before the antibiotic explosion and was a self-sustaining operation. Indeed, it had originally been the U.S. subsidiary of the German chemical and drug giant E. Merck, which had been seized during World War I and spun off into an independant operation. A few skeptics might have wondered if Merck or Lilly was overvalued, but it is hard to imagine them not making it.

9. Hybritech Inc. initial offering prospectus, Oct. 28, 1981, p. 11.

Chapter 9

1. Genetic Systems' initial offering prospectus, June 4, 1981.

2. Genetic Systems' 1981 annual report.

Notes

3. Equipment lease agreements between Genetic Systems and First Interstate Bank, exhibit 10–12, Dec. 28, 1981, March 22, 1982, and Sept. 9, 1982.

4. For a more complete explanation and analysis of burn rates in biotechnology, see three annual surveys published by the accounting and consulting firm of Arthur Young: *Biotech86: At the Crossroad*; *Biotech88: Into the Marketplace*; and *Biotech89: Commercialization*.

5. Burn rate estimates arrived at from publically disclosed figures published in Genetic Systems' secondary offering prospectus, April 7, 1983, and Genetic Systems' 1981 annual report. For history of the growing network of deals between Genetic Systems and other companies, see the secondary offering prospectus.

6. Genetic Systems' 1982 annual report.

7. Agreement between Genetic Systems and Syntex, May 19, 1981, exhibit 10–19.

8. Genetic Systems Respiratory Partners Research and Development Agreement, Dec. 31, 1981.

9. Genetic Systems' secondary offering prospectus, April 7, 1983, p. 16.

10. Burn rate estimates and quote are from figures in Genetic Systems' secondary offering prospectus, April 7, 1983.

11. James Glavin, interview with author, February 1987.

12. Agreement between Genetic Systems and Syntex.

13. James Glavin, interview with author.

14. George Todaro, interview with author, March 13, 1986.

15. Employment agreements with Dr. George J. Todaro, Nov. 10, 1982, exhibit 10–23.

Chapter 10

1. Information about Alfacell comes from a variety of public documents, research reports, press releases, and interviews. For instance, I consulted annual reports, proxy statements, 8-Ks and 10-Ks from 1984 to 1986—after which Alfacell's SEC reporting grows more intermittent. I also have in my possession a fairly complete collection of press releases from those years, as well as privately circulated research reports from Martin Blyseth dated April 4, 1984; April 5, 1984; Aug. 29, 1984; Jan. 1, 1985; March 10, 1985; April 4, 1985; Sept. 10, 1985; Oct. 15, 1985; Sept. 28, 1986; March 26, 1986. These were supplemented by a number of phone calls over these years with Blyseth that continued after he gave up writing the reports in 1987.

2. Pragma-Biotech's initial offering prospectus, Dec. 1986.

3. Alfacell's 1984 annual report, p. 3.

4. Ibid., p. 10, n. 9.

5. All quotations from Blyseth come from privately circulated reports in the possession of the author.

6. Alfacell's 1984 annual report, pp. 11–12, n. 10.

7. Pragma-Biotech's initial offering prospectus.

8. The chronology of events from 1985 onward can be followed in a series of Blyseth reports and in a series of corporate press releases.

9. Kuslima Shogen, from an Alfacell press release, Dec. 14, 1984.

10. *Professional Tape Reader*, March 22, 1985; *National OTC Stock Journal*, April 8, 1985; *Portfolio Letter*, April 22, 1985; *Barron's*, April 29, 1985.

11. The ticker story came from a press release from Alfacell's public-relations firm of Alan Bell & Co. dated May 9, 1985.

12. Kuslima Shogen, letter to shareholders, Sept. 22, 1986.

13. Alfacell lost its listing on the major over-the-counter market, NASDAQ, on November 5, 1986, when it failed to meet the association's capital requirements. The company, however, was still operating as late as March 1988.

14. Despite the extremely favorable treatment biotechnology won from the media, many in the industry regularly attacked it for not showing enough "support" in those years (particularly as stock prices fell), as if biotech were organized, like an academic lab, solely in the public interest. At an Industrial Biotechnology Association meeting in the

Notes

winter of 1984, a series of speakers was lined up to counsel the companies on press relations. Afterwards, I was speaking to an analyst who bitterly attacked a colleague who had risen to ask a few questions. "Why does he bother to come at all?" she snapped. "He obviously doesn't support the industry."

Chapter 11

1. Genetic Systems' secondary offering prospectus, April 7, 1983.
2. Nina Siegler, "Genetic Systems Corporation," PaineWebber, Jan. 3, 1983.
3. Nelson Schneider, E. F. Hutton Investment Summary, Nov. 28, 1983. Schneider also liked Genentech and Genex. Genentech, of course, was everyone's favorite, although there were the first murmurs that the stock price was too high. Genex almost collapsed in 1985 when G. D. Searle pulled out of a contract with Genex for producing one ingredient of the sweetener aspartame. Genex had already heavily invested in a manufacturing facility.
4. Quoted in William Pat Patterson, "The Jockey and the Mouse Doctor," *Industry Week*, Aug. 22, 1982, p. 42.
5. Sigiloff Ziering, interview with author, Oct. 22, 1986. All further quotations from Ziering in this chapter are from this interview.
6. Nelson Schneider, "Biotechnology Overview," E. F. Hutton, July 7, 1983, p. 5. One indication of the disparity in marketing forecasts comes from Nina Siegler's Paine Webber report on Genetic Systems, Jan. 3, 1983. Siegler estimates in that report that the immunodiagnostic portion of the total market is some 35 percent, $480 million out of $1.7 billion. That, of course, is seven times larger than Schneider's.
7. Peter Drake, "Monoclonal Antibodies and DNA Probes: Perspectives for Medical Diagnostics and Therapeutics," Kidder Peabody, April 2, 1984, p. 1.

Chapter 12

1. Hubert Schoemaker, interview with author, Oct. 3, 1986.
2. Wayne Fritzsche, interview with author, Dec. 16, 1986. The author had a number of other conversations with Fritzsche on this and other subjects from 1986 to 1988.
3. Quoted in "Bringing Major Corporation Savvy to a Biotech Pioneer," *Medical Business Journal*, Oct. 1985, p. 289.
4. Jack Schuler, interview with author, July 8, 1986. Unless otherwise noted, all further quotations from Schuler in this chapter are from this interview.
5. Phillip Whitcome, interviews with author, March 25, 1986, and April 19, 1986.
6. Transcripts of addresses to the Dean Witter Reynolds Inc. Health-Care Forum, April 13, 1983, p. 26.
7. Ibid.

Chapter 13

1. Grant Fjermedal, *Magic Bullets* (New York: Macmillan, 1984).
2. Transcripts of addresses to the Dean Witter Reynolds Inc. Health-Care Forum, April 13, 1983, p. 23.
3. Thomas McKearn, interview with author, Sept. 22, 1986.
4. Robert G. Mellem, "Genetic Systems," Piper, Jaffrey & Hopwood, Oct. 17, 1983, p. 1.

Notes

Chapter 14

1. Joseph Ashley, who became president in late 1984, downplays problems with the FDA. "There might have been some FDA delays," he says. "But that's a minor problem, an irritation. The real problem was that they didn't have a product" (interviews with author, April 12 and 28, 1987; May 19, 1987).
2. Robert Nowinski et al., "Monoclonal Antibodies for Diagnosis of Infectious Diseases in Humans," *Science* (Feb. 11, 1983): 637.
3. Janice LeCocq, "Genetic Systems Corporation," Montgomery Securities, June 13, 1983, p. 12. LeCocq admitted she did not expect products for two or three more years.
4. Bob Swanson at Genentech realized this early on. He not only hired bioprocess engineers long before there was anything to manufacture but also continually boasted of Genentech's manufacturing expertise publically. Indeed, when Genentech did begin to sell therapeutics, notably its clot dissolver, Activase, manufacturing the material in bulk, according to FDA specifications, did not prove to be a problem. Those dials Swanson was always posing before were, of course, from a fermenter, the heart of a biological manufacturing facility.
5. The patent application on Chemware was titled "Compounds and methods for preparing synthetic polymers that integrally contain polypeptides." By coincidence, the U.S. Patent Office granted a patent to the Swedish company Pharmacia that same day for a technique called "covalently binding biologically active substances to polymeric substances." Genetic Systems' patent attorney told *Biotechnology Newswatch*, an industry newsletter, that there was no conflict since its technique was in a more advanced state. The company received the patent in mid-1985.
6. Both Glavin and Nowinski are quoted in William Pat Patterson, "The Jockey and the Mouse Doctor," *Industry Week*, Aug. 22, 1982, p. 43. Nowinski also said: "I kind of drive Jim crazy because I've always got my fingers in everything. I'm a very detail oriented person" (p. 43).
7. Some scientific controversy still exists over whether the HTLV-3 virus is the causative agent of AIDS. Most scientists believe that it is, although a small minority of dissenters—notably Peter Duesberg of the University of California at Berkeley—have questioned that.
8. Robert Kupor, "Genetic Systems Company Update," Cable, Howse & Ragen, July 15, 1984.

Chapter 15

1. Bristol-Myers 1984 annual report.
2. Stephen Carter, interview with author, April 14, 1987.
3. Martin Kenney, *Biotechnology: The University-Industrial Complex* (New Haven: Yale University Press, 1986), p. 28.
4. Weinberg would soon after move his team into the Whitehead Institute, one of the new hybrid institutions formed between the university and an outside source of funding, in this case from John Whitehead, the founder of Technicon, a medical device company. MIT, of course, had traditionally had strong industrial ties. But Whitehead asked and received, after some controversy, control over the internal affairs of the institute whose researchers would retain a faculty position at MIT.
5. In its 1983 annual report, Genetic Systems boasted: "Scientists at Oncogen have discovered a new substance, oncostatin." Actually, Todaro's NCI lab announced the work in June 1982. Indeed, the researcher who physically did the work joined Stephenson at Oncogene Science.
6. George Todaro, interview with author, March 13, 1986.
7. The Hellströms got around. In 1983, as Todaro was hiring staff at Genetic Systems, both Hellströms were serving on Hybritech's scientific advisory board. Todaro got involved in a bit of a bidding war for the Hellströms with Imré, another small, Seattle

Notes

biotech startup. Imré was developing a medical device for removing certain immune complexes from blood. The heart of the system was a material called Protein B—discovered by the Hellströms. While Todaro eventually won the services of the pair, Imré managed to sign the Hellströms up as consultants.

8. James Glavin, interview with author, Feb. 1987.

Chapter 16

1. Genetic Systems 1983 and 1984 annual reports.

2. One of the clearest current analyses of the diagnostic business comes in a research report called "Demystifying the Diagnostic Industry," by Lorraine Schwarz of Saloman Brothers, dated Oct. 1988. The actual problems of marketing sexually transmitted disease tests came out in a series of conversations with health-care consultant Wayne Fritzsche.

3. Genetic Systems letter to shareholders, Dec. 19, 1984.

4. The same analyst visited with Nowinski in his office at dusk, as the sun sank over Puget Sound. The stereo played in the background, and Nowinski went on for some time "talking about all kinds of things, very philosophically. He was impressive, if a bit unorthodox." Later, the analyst scribbled: "Truly gifted, engaging individual."

5. Howard Teeter, interview with author, May 19, 1987.

6. James Glavin, interview with author, Feb. 1987.

7. Even as late as the second half of 1985, Todaro retained his freedom. Oncogen had always spent considerably more on patent applications than Genetic Systems. One reason, said Ken Gindroz, who became head of administration (which included patents) in June 1985: Todaro liked to use an outside attorney to prepare first drafts of patent applications, while Genetic Systems used a much cheaper, inside lawyer. It mounted up. At one point, Gindroz approached Todaro on the matter after Simmons had agreed that it made sense. "Not only did he [Todaro] not want to meet on it, he didn't want to discuss it at all," said Gindroz. " 'I like the way I'm doing it now,' he said and hung up. There were enough things to do at Genetic Systems not to get tangled up in this, so I dropped it" (interview with author, April 10, 1987). Ironically, Nowinski later gave Todaro oversight over the patent lawyer. That lasted only a short time before the function was ordered to report to corporate Bristol-Myers in New York, whose policy it was to use inside attorneys as much as possible.

Chapter 17

1. Joseph Ashley, interviews with author, April 12 and 28, 1987; May 19, 1987. All further quotations from Ashley in this chapter are from these interviews.

2. Udo Henseler, interview with author, April 27, 1987.

3. Kenneth Gindroz, interviews with author, April 10 and May 4, 1987.

4. The patent dispute was finally resolved in March 1987. Institute Pasteur and the U.S. Health and Human Services agreed to share the patent, while Gallo and Luc Mantagnier, with their colleagues, agreed to share inventor status. The agreement effectively eliminated Nowinski's dream of dominating the market, and it redirected the royalty stream to Pasteur into further AIDS research. It also resolved the rhetorical problems surrounding the AIDS virus: LAV and HTLV-3 were junked for HIV, or human immunodeficiency virus.

5. This wasn't completely Nowinski's fault. The early entrants in AIDS testing—Abbott and Electro-Nucleonics—had gotten approval from the FDA Bureau of Biologics in six months, four months faster than usual. (Because blood was injected into the body, it went to the Bureau of Biologics and required a longer approval process and more clinical testing than other *in vitro* tests.) The reason was the need for an AIDS test. Nowinski clearly thought that Genetic Systems would also be put on the fast track. But with a number of companies producing tests, the FDA clearly felt the crisis had passed,

Notes

and it returned to the longer cycle. And indeed it finally did take about ten months. Even fast-tracked, of course, the test would not have been approved until November, still wide of his forecast.

6. Hepatitis was the real key to the AIDS market. Until AIDS, Hepatitis A and B had been the only virus systematically screened for among the blood banks, a market Abbott had long controlled through the Quantum. When Abbott came out with its tests, it was relatively simple and inexpensive for the blood banks to add it to the Quantum system. Electro-Nucleonics, a New Jersey company that came out a few days later with its AIDS test, had, on the other hand, always done well in the smaller market of blood fractionaters, made up of organizations that break down the blood into its component parts. After approval, it continued to do well in that market, although Abbott took the bulk of sales, up to 70 percent.

7. Chemware was eventually replaced by an ELISA format, a system using enzymes to amplify the signal. In that case, a company called Organon, the developer of the ELISA format, received a royalty. Genetic Systems also had to pay a royalty to the federal government for the use of a cell line used in the manufacturing process.

8. Robert Nowinski, interview with author, Oct. 17, 1985. This interview was conducted just a few days before the company was sold—not to Syntex, but to Bristol-Myers, while the author was at *Forbes* magazine.

Chapter 18

1. The situation was actually more complex than that. Johnson & Johnson had already established a web of relationships with other, second-generation antibody companies such as Immunomedics. The three first-generation companies were not only the most mature, but, relatively speaking, were the least constrained contractually.

2. Howard Greene, interview with author, Oct. 1985, as preparation for "Fatal Flaws?" *Forbes*, Nov. 18, 1985.

3. Tandem provided continuing legal work for Hybritech lawyers. The system was widely adopted throughout the industry, despite a patent Hybritech won in March 1983. Rather than tangle with Abbott, Hybritech struck back at another small California company, Monoclonal Antibodies, which used it in its ovulation test. The case dragged out for several years, with several startling reversals of fortune. Monoclonal won the first few rounds when a California judge decided that there was enough "prior art" floating around to overturn the patent. However, in 1986, an appeals judge reversed the judgment, awarding the case to Hybritech. By that time Hybritech had been acquired by Eli Lilly. Now, with Lilly's cash and legal staff on hand, Hybritech turned on its old nemesis, Abbott, which was extensively using Tandem for a series of increasingly successful physician's office tests. This issue, as well as the Hybritech antitrust suit, continues to rattle around the court system.

4. All this was relative. Hybritech had sold the marketing rights to products for the over-the-counter market to an unnamed company for $5 million. It gave the rights to the physician's office tests to Access Medical Systems, in which it had an equity stake. Curtis Matheson distributed its clinical laboratory products. Other companies had overseas rights. Other deals were signed for future therapeutics products.

5. Ted Greene objected strenuously to the notion that the business plan based on leveraging from diagnostics to therapeutics was fundamentally flawed. He argued in an unpublished letter at the time (in response to "Fatal Flaws?" by Robert Teitelman, *Forbes*, Nov. 18, 1985, p. 94) that "opportunities for investing in development of high potential drug products now exceeds our early expectations." Three years later, however, no former Hybritech top manager remained at the firm, including Greene, and Lilly had had very little success with anticancer antibodies, despite the expertise of Hybritech, the nearby Scripps Clinic, and its own extensive in-house research group. Indeed, the one antibody that Lilly briefly expressed enthusiasm for—called KS1/4, against lung cancer—was a Scripps development. So far it has not panned out. Dr. Order continues his work at Johns Hopkins.

6. An additional factor was the length of time the early investors—the Kleiner, Perkins

Notes

group, for instance—had left their money in Hybritech. Eight years is a long time to park a venture investment, even if profits beckon. Thus, the situation: a treacherous environment, a good year in terms of profit, investors anxious to take home profits, and management needing large amounts of cash to fund therapeutics.

7. Joseph Ashley, interviews with author, April 12 and 28, 1987; May 19, 1987. All further quotations from Ashley in this chapter are from these interviews.

8. The CPU was very similar to General Motors' purchase of Ross Perot's Electronic Data System a year earlier. Like the CPU, GM set up a new class of E stock linked to EDS performance. The rationale, again, was to preserve the entrepreneurial spirit of an organization by giving it an incentive separate from the company at large. GM later set up another class of F stock when it bought Hughes Aircraft. For a good technical discussion of these sorts of deals, including EDS and Hybritech, see Wayne Pambianchi, "Earn-out Mechanisms Allow Buyers to Retain Sellers' Commitment," *Medical Business Journal,* Oct. 1985, p. 296.

9. Peter Drake, interview with author, Oct. 16, 1985, and phone conversations, 1985 to 1987.

10. Syntex was not about to watch a major part of its diagnostics future slip from its grasp without getting something in return. A settlement was finally struck. Syntex received $2.45 million from Genetic Systems for funds advanced on various projects and some $15 million for its one-third share in Oncogen, a considerable premium over what it had actually spent. If Oncogen should commercialize certain projects, Syntex could get up to another $10 million more. In exchange, Syntex agreed to cancel the 1985 diagnostics deal, although it retained the Microtrak tests. It continued to hold 2 percent of Genetic Systems stock, a last, lingering, faintly mocking reminder of that first deal when the Blechs had sold it stock at one dollar a share.

11. Kenneth Gindroz, interview with author, April 10, 1987 and May 4, 1987.

Chapter 19

1. Among the group was the author and Dr. Ingegard Hellström of Genetic Systems. Rumors were already swirling of some sort of buyout, although they centered on Syntex, not Bristol-Myers.

2. Gene Bylinsky, "A New Cancer Breakthrough," *Fortune,* Nov. 25, 1985.

3. Peter Drake, "Product Asset Valuation: Benchmark for Valuation in Biotechnology," Kidder Peabody, April 11, 1986.

4. Stelios Papadopoulos, "A View from the Cell Side," Donaldson, Lufkin & Jenrette, March 21, 1986.

5. Ibid, p. 7. The author conducted numerous other interviews with Papadopoulos from 1986 to 1989 on various aspects of biotechnology.

6. Stelios Papadopoulos, "Biotechnology Monthly," Drexel Burnham Lambert, March 1987, p. 6.

7. The price was an issue of some debate. Activase sold at $2,200, quite a bit higher than streptokinase, which went for $200. Early in 1988, Medicare refused to increase its reimbursement rate for coronary patients just to account for Activase's high price. This too may have played a role in dampening demand.

8. The stock steadily slid, from 53 just before the Oct. 19 crash to the mid-teens at the end of 1988. Based on 1988 earnings of $.60 a share, the market was ratcheting Genentech down from a company that deserved a price to earnings multiple of 88 to, at a price of 15, one of 20. At the same time, the average drug p/e was around 16, while the gold-standard Merck was getting 24.

9. The author discussed the Nova deal and several others in "What a Sale: Biotech's Bargain Basement Arrives," *Oncology Times,* Aug. 1988, p. 5.

10. See Roger Longman, "Equal Partners," *In Vivo,* published by the Wilkerson Group, Sept./Oct. 1988, p. 25.

11. The research boutique is a perfectly viable form of business organization. See Robert Teitelman, "Biotech Boutiques May Lead to Greater Financial Rewards," *Oncology Times,* March 1989.

Notes

Chapter 20

1. Thomas Kuhn, *The Structure of Scientific Revolutions* (Chicago: University of Chicago Press, 1962).

2. Relevant here is a series of essays entitled *Infinite in All Directions* by physicist Freeman Dyson (New York: Harper & Row, 1988). Freeman divides the sciences into two categories: those seeking a single, overriding absolute to nature, what he calls the Athenian, or those taking refuge in the sheer diversity and complexity of life, the Manchester school. Twentieth-century physics, of course, which molecular biology has modeled itself after, fits the Athenian ideal. Dyson, on the other hand, although a physicist, throws his lot in with the Mancunians—named for the role Manchester played in the early Industrial Revolution—the empiricists, the fiddlers and inventors. "Life by its very nature is resistant to simplification, whether on the level of single cells or ecological systems or human systems," writes Dyson in one of the essays, "Why Is Life So Complicated?" (p. 95). For a further elaboration into Dyson's dichotomy, see Stephen Jay Gould, "Mighty Manchester," *New York Review of Books*, Oct. 27, 1988, p. 32.

3. See Robert Teitelman, "The Baffling Standoff in Cancer Research," *Forbes*, July 15, 1985, p. 110, for Duesberg's and Rubin's view on oncogenes. On the AIDS question, see *Science*, (July 29, 1988): 514, in which Duesberg debates the question with NCI's Robert Gallo and William Blattner and Howard Temin at the University of Wisconsin.

4. Harry Rubin, "Is Somatic Mutation the Major Mechanism of Malignant Transformation?" *Journal of the National Cancer Institute* (May 1980): 999.

5. Harry Rubin, interview with author, Sept. 1986. All further quotations from Rubin in this chapter are from this interview.

6. The issue of cancer statistics is a contentious one. Over the years a number of studies have attacked NCI's position that cancer mortality figures are improving. In 1985 a group of Harvard researchers went public with doubts in *Scientific American*. Two years later an even more damaging report appeared from the federal government's General Accounting Office. Both studies questioned the usefulness of NCI's five-year survival benchmark—if you survive five years, you are considered cured. The reason: With earlier diagnosis, more cancer patients survive five years, although approximately the same percentage eventually succumb to the disease. In reply, the head of NCI, Dr. Vincent DaVita, argued that the GAO report ignored "the enormous progress" made in understanding the cancer cell. In a written rebuttal, the federal Health and Human Services attacked the study as "opinion, not fact," calling its tone "negative" and "counterproductive." Here was a battle not only over research initiatives and statistics, but over funding, and fundamentally, between the forces of optimism and pessimism.

7. Susan Sontag, in her essay about cancer, *Illness as Metaphor* (New York: Vintage Books, 1977), offers her own opinion on the division between optimism and pessimism in cancer research: "More recently, the fight against cancer has sounded like a colonial war—with similarly vast appropriations of government money—and in a decade when colonial wars haven't gone too well, this militarized rhetoric seems to be backfiring. Pessimism among doctors about the efficacy of treatment is growing, in spite of the strong advances in chemotherapy and immunotherapy made since 1970. Reporters covering 'the war on cancer' frequently caution the public to distinguish between official fictions and harsh facts; a few years ago, one science writer found American Cancer Society proclamations that cancer is curable and progress has been made 'reminiscent of Vietnam optimism before the deluge.' Still, it is one thing to be skeptical about the rhetoric that surrounds cancer, another to give support to many uninformed doctors who insist that no significant progress in treatment has been made, and that cancer is not really curable" (p. 65).

8. See Natalie Angier, *Natural Obsessions: The Search for the Oncogene* (New York: Houghton Mifflin, 1988). Angier's book focuses on Whitehead Institute lab chief Robert Weinberg. Of particular note: Weinberg articulates clearly his belief that the answer to cancer is simple and elegant and lies in the genes.

9. See, for example, Ted Howard and Jeremy Rifkin: "For years social commentators have looked on nuclear weaponry as the most powerful and dangerous tool at the disposal

Notes

of humanity. With the development of human genetic engineering, a tool even more awesome is now available" (*Who Should Play God?* [New York: Dell, 1977], p. 9).

10. See Jesse Treu, "Biotechnology Seems to Follow Semiconductor's Route as It Looks Beyond 'One-for-One' Products," *Medical Business Journal,* July 31, 1988, p. 218. Treu offers a more sophisticated, three-stage model of the semiconductor analogy which takes into account the fact that early biotech offerings like human insulin and t-PA seem to be more commodity than value-added products. Indeed, Treu dismisses the first generation of companies and turns his attention to the next. I offered a counterargument in a two-part essay in *Oncology Times,* Oct. 1988 and Nov. 1988.

11. See Robert Johnston and Christopher G. Edwards, *Entreprenurial Science: New Links Between Corporations, Government and Science* (New York: Quorum Books, 1987) for one of the balder statements of this view. The authors argue not only that technological startups will save the country—"Expect that high technology will be America's economic panacea" (p. 2)—but that the government should help them along in any way it can. They also make an argument exactly contrary to Harry Rubin about the maturity of the technology. "One important characteristic of biotechnology is the very short lead time from discovery to application. A laboratory finding can, in many cases, lead to a path of product development almost immediately after the finding is published" (p. 7). This view seems overly optimistic. Robert Johnston, by the way, is the same venture capitalist behind Cytogen and Genex.

12. The transistor flowed from communications technology, hence its link to Bell Laboratories. Ernest Braun and Stuart MacDonald, in their study *Revolution in Miniature: The History and Impact of the Semiconductor Revolution* (Cambridge: Cambridge University Press, 1978) go back to Michael Faraday's discovery of electromagnetic induction in 1833 as the genesis of microelectronics. The figures on transistors come from them as well (pp. 54–55).

13. See *Science,* March 18, 1988, p. 1364, and March 25, 1988, p. 1979, for a comprehensive examination of the problems at NIH.

Chapter 21

1. Stephen Carter, interview with author, April 14, 1988. All further quotations from Carter in this chapter are from this interview.

2. Carter was, if anything, even more emphatic about cytotoxics at an analysts' meeting held the month before, in March 1988: "Why do we continue our activities in cytotoxic chemotherapy? . . . First, it's a mainstay of our cancer business. Second, and perhaps most important, no biological therapy currently in clinical trial has demonstrated a major impact on the survival of cancer patients. It is clear that biological therapy today, at least those discoveries that are in the clinic, do not threaten the use of chemotherapy. If anything, the proponents of biologic therapy are moving toward the concept of combining the two approaches. The third reason is that we feel cytotoxic chemotherapy is most likely to be an important modality into the early twenty-first century."

3. An insight into Todaro's views comes from that same March 1988 analysts' meeting: "The last one and a half years as part of Bristol-Myers have been exciting because we've been able to concentrate on new discoveries and on bringing them forward. We have maintained a high level of scientific competence and, except for today, have not had to concern ourselves with financial analysts. If you knew the previous two or three years at Genetic Systems, that's quite a change—and a beneficial one."

4. Ironically, the word *revolution* originally came into use in the physical sciences. As Harvard historian I. Bernard Cohen points out in his study *Revolution in Science* (Cambridge: Belknap, 1985), the term was first used to denote a physical rearrangement of the universe, as in *De Revolutionibus* by Copernicus in 1543. With the coming of the Enlightenment, the term became synonymous with massive political and social change, particularly the French Revolution, before seeping back into the sciences in the nine-

Notes

teenth century. In this case, we are not talking about its use in various fields but rather its promiscuous, dilutive use.

5. In the words of Daniel Bell in *The Cultural Contradictions of Capitalism* (New York: Basic Books, 1976): "In two fundamental ways the new revolution has already begun. First, the autonomy of culture, achieved in art, now begins to pass over into the arena of life. . . . Anything permitted in art is permitted in life. Second, the life style once practiced by a small *cenacle* . . . is now copied by the 'many' . . ." (p. 53).

INDEX

Index

Index

Index

Index

Index

Index

Synergen, 192

Syntex, 44–45, 56, 71–72, 142; and Genetic Systems, 88–90, 171–74, 177–78, 181, 196; and Oncogen, 162, 163; *see also* Syva

Syva, 44, 74, 84, 87, 106; and the Emit system, 135; and Genetic Systems, 119, 120–21, 171–83; and immunoassays, 117; *see also* Syntex

Szebenyi, Andrew, 94

Szebenyi, Emil, 93, 95, 96, 97

Szebenyi, Stephen, 94

Takeovers, 78–79, 174–185, 186, 190

Tandem, 23, 176, 226n3

Tax shelters, 85, 89, 160

TDx, 119, 121, 135

Teeter, Howard, 161

Temin, Howard, 59

Texas Instruments, 115, 116–17

TGFs (transforming growth factors), 64–67, 152, 213

Thornton, Dean, 168

3T3 cells, 56, 57, 58, 62, 66, 150

Time (magazine), 12, 31, 32, 176, 186

Tissue-plasminogen activator (t-PA), 9, 193, 198

Tobacco Institute, 38

Todaro, George J., 52, 54–70, 106, 119, 200, 210; and the buy-out of Genetic Systems, 183–84; and oncogenes, 148, 149, 151–54, 163, 183

"Today" show, 189

Transfections, 67

Tuberculosis, 16

Tumors, 33, 56, 66, 93–97, 102; and anti-cancer drugs, 145, 146; and the naked magic bullet theory, 123–24; and oncogenes, 149

United Artists, 30

United Nations, 29

Upton, Arthur, 68

Vaccines, 24

Vagelos, Roy, 207

Valuation, 190–91

Varmus, Harold, 61

Velcro, 34

Vilcek, Jan, 31, 35

Vincent, James, 114, 115–17

Viral Carcinogenesis Laboratory, 57, 60, 68

Virgilio, Abramo, 163, 184

Viruses, 28–29, 52–61, 62–63

Wall, Michael, 125, 184

Wall Street Journal, 189–90

War on Cancer, 8–9; funding for, 17, 21, 67; Lasker and, 30; legislation, 14, 30; Rauscher and, 32; *see also* Cancer cures

Warhol, Andy, 210, 215

Warner Lambert, 119

Watson, James, 15, 18, 51–53, 54, 61, 206

Weinberg, Robert, 149, 150–51

Weissman, Charles, 60

Weissmann, Gerald, 34

Whitehead, John, 13

Whitman, Walt, 54

World War II, 14, 16, 141, 142, 206

Xeroxes, 12

Xoma, 198

Ziering, Sigiloff, 109

Zinder, Norton, 81